12 RULES FOR PRO CODERS

ACTIONABLE ADVICE TO ACCELERATE YOUR CAREER

GREG L. TURNQUIST

MOUNTAIN
SUMMIT PRESS

No part of this manuscript was produced using ChatGPT, DALL-E, or any other AI tools. Such tools may have been used for copywriting purposes such as titles, back cover copy, and advertising campaigns.

© Cover Design by Cora Graphics

© Photo by Depositphotos.com

Greg L. Turnquist

c/o 12 Rules for Pro Coders

P.O. Box 4042

Clarksville, TN 37043

USA

ISBN: 978-1-956410-39-6 (ebook)

ISBN: 978-1-956410-37-2 (paperback)

ISBN: 978-1-956410-38-9 (hardback)

This book is dedicated to those who help others learn to build, grow, and develop their skills, whether they are professors, advocates, trainers, or educators. You truly make us all become pro coders.

This book would not have been possible without these
12 RULES ULTIMATE FANS:

Donald Coffin

CONTENTS_

PREFACE_

Where was this book when I was fresh out of college? I knew C, C++, Java, and enough Perl to be dangerous. IDEs were new then, and I had never used one before graduating college. I had never heard of version control or design patterns, but I believed I was ready for my first coding job.

A defense contractor hired me to rewrite some radar processing code for a supercomputer. The previous author didn't believe in documentation, and no one handed me a test case to ensure this software worked. The project was written entirely in C and didn't include a Makefile, so I was left to discover what libraries were used to compile the project. I only understood two things: code and math, but nothing about radar processing theory. I translated each line, one by one, to the MPI C library with no idea how radar antennas actually worked. A couple of months into the project, they gave us our first test data sets, and we proved the software worked and brought processing time for a single image frame from minutes to under one second. The company was overjoyed that I got this to work, and it generated a flurry of activity. The company invited our

Congressman to tour our offices while they schmoozed him for money. You'd think this was a success story of overcoming a difficult circumstance, but it isn't.

I failed spectacularly at my first job because I lacked a professional persona. I was young and immature, and it showed. The incoming project didn't have documentation, so I didn't bother writing any documentation either. I was let go shortly after having completed the project. Devastated after losing my first job, I became an adjunct instructor, which eventually led to becoming a Computer Science professor. I now have the privilege of speaking regularly to classrooms of students, all hoping to be software engineers someday.

Being a software engineer carries with it so much more responsibility than just writing code, and that's a lesson I learned the hard way. It means being a salesperson for yourself, conscious about the impact of solutions you are crafting, and a team player for your fellow engineers, managers, and stakeholders. If I could sum up everything that is part of being a professional coder that doesn't involve coding in a single phrase, it would be "responsibility to the profession." This responsibility includes documenting, testing, and continuously integrating your code, but it also includes managing your time, attention span, meetings, and those hired after you.

This book outlines the rarely-spoken essentials that few coding books or college classrooms cover. I've gotten to know Greg and his work with Spring.io over the last several months. Greg creates engaging videos and lectures, and that engagement made this book a joy to read. I've come to appreciate that Greg is being proactive about wanting college students and coding boot camp attendees to be better prepared for the rigorous demands of the coding profession.

You don't have to be overwhelmed by the coding profession, but you do need to be prepared, and you've come to the right place.

Dr. James Church
Associate Professor of Computer Science
Austin Peay State University

RULE 1: CURATE A PROFESSIONAL PERSONA_

To become a pro coder, you must embrace several things. While this may sound like a job or the beginning of a career, this particular industry has a unique nature that extends way beyond just coding.

There is more to this field than simply writing code, testing, sustainment, training, and helping others solve problems. A significant aspect we are entering today is the online cyber era. We aren't just reporting for work at some job site. We must prepare to interact with an entire world of other developers, customers, and community members that span the globe.

In this book, you will learn a collection of actionable rules to accelerate your career. While no one rule is the be-all and end-all for everything it means to be a pro coder, embracing these rules will help you make the leap from junior to senior.

And to do this properly, we must embrace ***Rule 1*: *Curate a Professional Persona***.

So what is a persona?

It's your personal brand. The way you appear to others. While you may interact with most people face-to-face on a daily basis, you'll also deal with them through other means during your career including email, web, social media, and source code management. All of these are avenues of communication, and you must be sure to build and preserve your personal brand across all channels.

While you could sculpt a profile that will support you interacting with fellow coders on development endeavors, it's a lost opportunity if you don't assemble the right parts and craft a persona that inspires trust amongst teammates, encouragement from managers, and hope from users.

So let's check out each section below to design and develop your developer persona.

One of the most important things you should have is a **professional** profile. And today there is no better place than **LinkedIn**.

LinkedIn provides the means to enter professional basics including, name, location, education, training, skills, experience, and track record.

That list may feel slim, but it's actually quite powerful. Let's dig in and find out why.

Personal identification is a bit obvious. But what's also key is providing the means to turn your education, training, and certifications into a curated list. Whether you're shopping for a standard salaried position at a tech firm or seeking clients for your consulting business, you should have a nicely polished set of credentials. This applies whether you complete a 4-year university curriculum, a 12-week boot camp program, get your Ph.D. or M.S. in Computer Science, or knock out a half dozen certifications. All of these are valuable credits you should gather. And LinkedIn makes it super simple to list them.

And that's a critical step.

Showing off your skills and official recognition is valuable.

Now it's been shown that pedigree isn't the *sole* factor in determining someone's skill or talents, but don't pass up the opportunity to cite the fact that you completed both a B.S. and M.S. in Computer Engineering at your university (if you did so).

But the next critical step in curating your LinkedIn profile is to start recording your work experience. Certainly cataloging the day you start a new position at a given company is a must. After all, that's LinkedIn's core feature. In your LinkedIn profile's **experience** section you have the opportunity to list the title given to you by whatever outfit you work for.

However, don't hesitate to *also* list other experiences that may fall outside of your official title. As an example, I joined VMware back in 2010 as a part of the Spring team. While the Spring team floated between VMware and Pivotal Software in a somewhat turbulent corporate ride, my commitment to that team never wavered. Hence, I have a *separate* entry documenting when I joined the Spring team, and when I switched roles on that team.

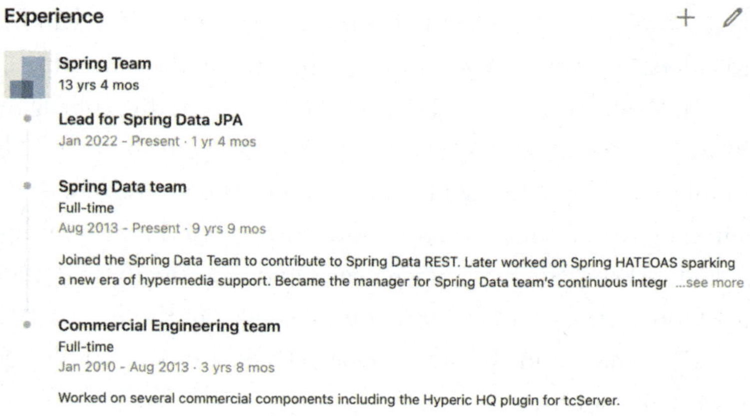

If you are contributing to a particular open source project in your spare time, don't hesitate to capture that as another experience. While you don't have to capture *everything* that happens on that project, it doesn't hurt to cast a more detailed picture of all the pursuits you are involved in from a professional point of view.

"But Greg, what about creating my professional bio on my *own* website?"

Yes, you can do that. And I even believe you can make it look nice, perhaps nicer than LinkedIn can. I have confidence in your ability to craft stylish CSS, but you shouldn't do that.

What? Why?!?

Because it's a waste of your time.

Virtually everything throughout this book can be done by you, by hand, in your own special way. And if you take the time to add up *all* the effort to do it "your way," you'll find that this level of effort isn't useful. There are, in fact, better ways to invest your spare time, which we'll cover later in this book.

But there's one other aspect to not doing it yourself—LinkedIn is also a professional social network. It's a place where you can build up interactions with other professionals in your field. You can post links to blog posts you'll write. You can link to videos you create. You can write directly on the platform.

And all of this can be seen and enjoyed by colleagues, hiring professionals, and others. All for your benefit. The number of people that will spot this on your own website is way slimmer, unless you *actually* know how to build a heavily-trafficked website. Such a task isn't easy, and as I already said, there are more effective ways to get your message out there.

But what about GitHub? Doesn't GitHub let you also get your name out there?

I have seen many arguments how GitHub, the source code repository site, is a natural fit for capturing your persona. After all, doesn't it show you "in action" as you work on various projects?

Simply put, GitHub fails on several fronts as being your professional "record," and here's why.

GITHUB IS TOO LOW-LEVEL

No recruiter or potential client has time to wade through your treasure trove of code changes and closed tickets. These people want a one page executive summary of you, something they can scan in eight seconds to decide if they should dig further.

And no, I'm not exaggerating about there only being eight seconds. Back in 2018, Ladders Inc. conducted a study and reported the recruiters, on average, will skim a resume in 7.4 seconds. Eight seconds is actually rounding things up.

GITHUB ISN'T BUSINESS-FRIENDLY

GitHub's focus is managing source code. Any sort of visibility or user profile is a secondary feature. While you can certainly leverage it (and we'll cover exactly the right way to do this, further down in this chapter), it won't be a first stop for hiring managers or clients, i.e. the people with money who can offer you a role. Therefore, it shouldn't be *your* first step for building a professional persona.

GITHUB IS UNCURATED

Most of the "action" that GitHub tracks is everything you touch from a source code perspective. The GitHub "wallpaper" of activity (of which some funny memes have been generated) isn't what you think it is. Look at this example below from my own profile:

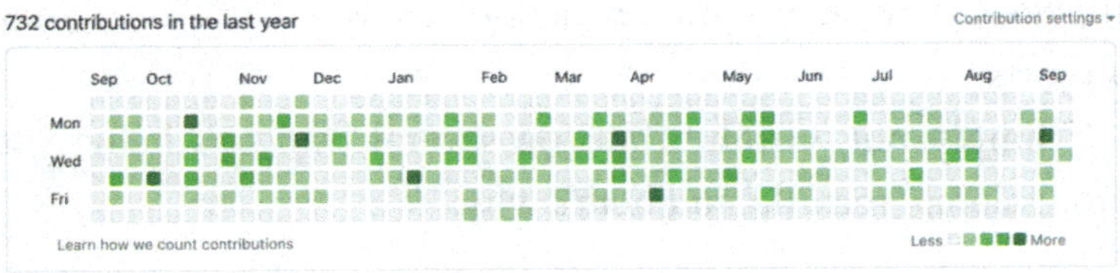

That dynamically rendered graphic is generated on a daily basis, based on your actions. It is NOT something to serve up to a recruiter. At a glance, it's pure noise, not something that whets their appetite and makes them want to dig deeper.

Instead, you want to give a potential recruiter something where your best efforts are shown first, not the results of a script that brute-forced an update to sixty-five different GitHub projects.

NOTE: I once had a script for just that purpose. My "activity" was ridiculous on GitHub back then!

Furthermore, you shouldn't be crafting commit messages based upon how a recruiter will view them. You certainly shouldn't be commenting on tickets wondering if a future hiring manager will enjoy your ticket-managing skills. Instead, write commits so that fellow contributors can read the record and understand how the codebase evolved. We'll cover this in more detail in ***Rule 5: Structure Matters***.

When it comes to your one-page executive summary and the expansion thereof, use LinkedIn as the way to curate what people need to read at first glance. That is what it's built for.

GITHUB DOESN'T TELL THE WHOLE STORY

The first half of my career didn't involve a single line of open source code. For thirteen years, the team I contributed to used private tools and kept the source code in-house. Perhaps this is your story too? It further illustrates how GitHub may not tell the whole story about you.

It doesn't mean you're doing something wrong. Or that your efforts are subpar in any way. And the fact that my contributions were not visible didn't stop me from securing a position on the Spring team.

That's because I curated a persona that made me irresistible. No, I didn't use LinkedIn, because at the time, LinkedIn was a startup and somewhat clunky. The LinkedIn of today is different, hence the reason I recommend it today. Nevertheless, I deployed many of the similar tactics we're covering in this chapter!

> WARNING: If you go hunting for blog posts I wrote years ago, you will find personal rants about the lack of value I found in LinkedIn back in the day. And I stand by those criticisms. However, LinkedIn has encountered a rebirth. It has become a social media platform for pros. You can easily craft a listing of positions, teams, education, and training. And you can arrange it as desired. In fact, it was enough that I re-opened what had been a closed account and came back. Here I am recommending it to you. It's truly worth your time.

I suggest you open a LinkedIn account. Whether you are in the process of completing some university program or are engaged in a boot camp curriculum, go ahead and start fleshing out the thing you are involved with in the educational section of your profile.

Alas, that is not the only thing you must do in building your professional persona. And so let's go over this checklist of things you need in your LinkedIn profile:

1. CHOOSE YOUR TOP-LEVEL DESCRIPTION

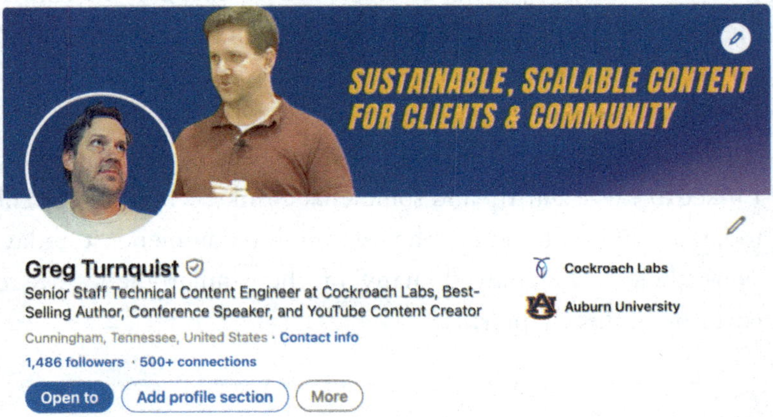

In my own profile, right under my name, you can see my particular work title "Senior Staff Technical Content Engineer" combined with the relevant group, "Cockroach Labs." Since I also began making technical content as of 2019, I also include "YouTube Content Creator." There are some other parts in there such as the fact that I've written multiple books and have spoken at many conferences.

If you're not certain what to put, at least start with your assigned title from your position. If you are still training or at school, you can choose something cute but indicative, like "Aspiring TypeScript coder" or whatever you feel fits you best.

And finally, you'll notice that I actually built a banner

graphic as well as a profile shot, both with a signature color scheme. Now I use Canva for doing graphics. You don't have to go that far, but you may want to give it a little thought.

2. START WORKING ON YOUR "ABOUT" SECTION.

The "About" part of your profile is a fast summary. It's where you want to list the most important things you'd want any recruiter to know. This is also known as a hook so they want to read more of your profile.

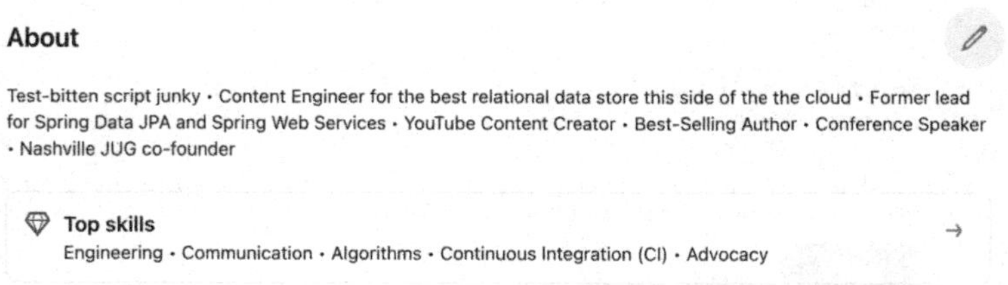

This is mine. My tagline has been "test-bitten script junky" for years. Try to think of a single sentence that captures your fundamental vibe about coding. If you're not sure, don't be afraid to start off with something and refine over months or even years.

There are also a list of my top skills I want recruiters to be aware of. It's possible to continue listing skills further down in your profile, but I frankly find that TOO much of that clutters things up.

The point is to start thinking about how you want others to see you.

By the way, while this part of LinkedIn isn't exactly flashy, it's not supposed to be. This is when you move past "stop the scroll" graphics and start providing real, concrete details that you want others to see.

3. FILL IN YOUR EXPERIENCE DETAILS.

Experience lists your actual roles and accomplishments. While it's primarily about focusing on companies and official teams, it's not confined to that. I already showed the Spring team bits earlier. Here's the rest:

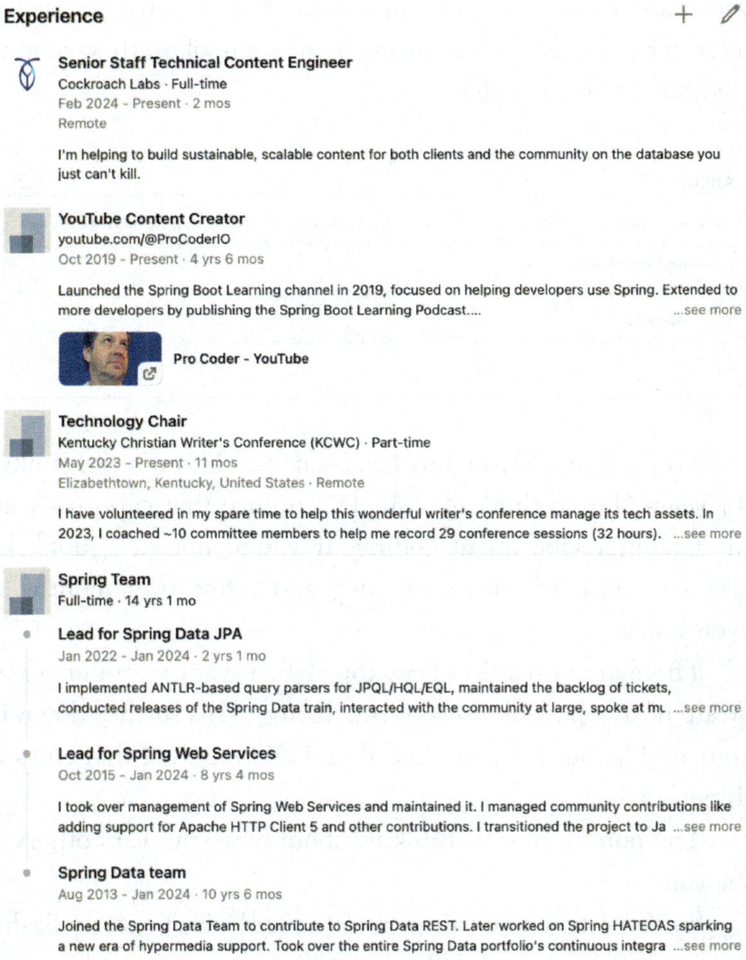

This list nicely goes backward in time, back to older positions. Mine goes all the way back to 1997 with my first role (not all is shown above).

And I created a completely separate entry for my YouTube efforts including a link to the channel so it won't be hard to find for anybody that's interested. Once again, the text is up to you. Instead of writing down absolutely everything that happened, use them as opportunities to highlight the most important aspect of any existing entry.

> WARNING: If at all possible, make sure the timeline is complete with no gaps. If you visit my profile on LinkedIn and scroll all the way to the bottom, you would see things running all the way back to December 1997. Be warned that there are people that look for such gaps and may try to infer something sinister about it. Don't give them the opportunity!

---

4. LIST ANY EDUCATION OR CERTIFICATIONS YOU HAVE.

Education

Auburn University
Master of Science - MS, Computer Engineering
Sep 1995 - Dec 1997

Auburn University
Bachelor of Science - BS, Computer Engineering
Sep 1991 - Sep 1995

List any degrees you have, any bootcamp programs you've completed, and even list your certifications. It's important to understand that some shops won't give you the time of day if you don't have a Computer Science degree while others don't care. Some are more interested in certifications than others.

It's important to flesh out these details for one key reason: if some hiring manager is trying to pick between you and another candidate, having every valid credential on record could be the trick to getting an offer.

Now there are other things you can list including publications, patents, languages you speak, and organizations you are a part of, but it's important to understand that the further you go down the list, the less important that section is. By all means, fill things with proper detail, but the places that warrant the most attention to detail are toward the top.

Also, don't sweat it if you don't have everything polished on Day 1. It can take time to fine-tune things. Just consider it something to revisit at least once or twice a year.

But that is not all.

The next thing to pay heed to...is your GitHub account. Maybe that's confusing because earlier in this chapter I ranted against GitHub in no uncertain terms. We'll cover why you actually *do* need a GitHub account, and how to best use it in the next section.

IT's true that GitHub is not your resume. It's also true that GitHub is built for source code management and not for "hanging your shingle" out there for people to find you. Nevertheless, GitHub is vital to your career as a pro coder to the point that you *need* a GitHub account.

So what are we talking about?

GitHub is the true "hub" of source code action. Most open source projects are there. And even many closed-source projects are *also* there. For this reason, it makes sense for you to be there as well.

And not just because source code is housed for countless projects. The fact that GitHub is used by teams around the world to manage tickets, feature upgrades, project planning, and other aspects of source code means that you simply can't avoid this platform.

It's a natural fit for you to open a free account at GitHub.

Even if your team is using something else, odds are pretty strong that *at least one* of the tools you will be using will have a presence on GitHub, and if you need to open a ticket to report a bug, it will be on GitHub.

But that is not all.

Inevitably you'll make a commit on some project that another developer or recruiter will happen to stumble across. And they'll look you up!

Why pass up the opportunity to give them a glance at who you truly are?

To be clear, this is not your primary presence, but simply an extension. Think of it as "Me as I appear on GitHub."

TIP: If at all possible, it's easier if you can grab the same handle *everywhere*. I own "gregturn", but only because that happened to be my original handle way back in the 90s. (Ancient, right?) I've grabbed that on GitHub (https://github.com/gregturn), Twitter (or X??) (https://twitter.com/gregturn), and I once had that handle on YouTube (that changed for other reasons). If possible, consider having a consistent "handle" on all your platforms.

Once you've set up your GitHub account, you have an opportunity to setup a simple profile that can serve you for the long run.

1. Put your name and a nice headshot; ask a friend to help shoot your first headshot using your phone.
2. Put the critical details in your profile description, including any titles associated with your position and your key responsibility to your team.
3. Craft a summary in a README.md file of who you are, what you do, and how people can follow you.

This is where you get to use the magical marketing tool known as "reused content." If you already figured out your "tagline" in the previous section for your LinkedIn profile, you can use it again right here!

Now I'm going to share one additional tip that not everyone knows, so listen closely.

Anytime someone stops on your GitHub profile or some other blog post you've written, or anything else you own on the Internet, always always *always* give them one more place to go. In short, ask yourself, "If someone were to visit my GitHub profile, what else would I want them to see?"

Because I make videos on YouTube that are related to my career, I always try to link one of my YouTube videos, the one ideal to the situation. Check it out below:

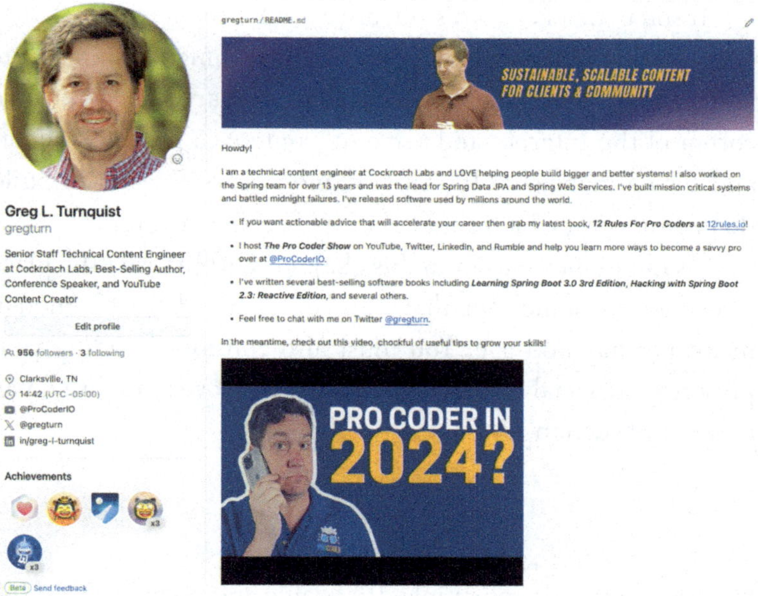

For starters, you can see my title, company, key responsibility, and other details on the left-hand side. Next, you can see more detail on the right.

I've written multiple software books, and I have links to them. I also have a link to a video I shot back in the fall of 2023 called "How to be a pro coder in 2024" to show tips I handed out to a classroom of computer science students at the local university. I used to list a different video which got lots of *traffic*. I hope to repeat the pattern!

If you have nothing else, then consider including a link like "Be sure to check out my LinkedIn profile and I'll be happy to connect over there!"

Now carry on the thought. If someone lands on your LinkedIn page, try to figure out what you'd want someone to do from there. And if you write a blog article in the future, you can always finish the article with a link to yet another article or video you've created.

If someone likes your stuff, always offer them more.

Yes, this is a bit of a content creator thing, pushing more and more content. But don't be afraid to carve out your own corner of the Internet and serve up content to potential clients and hiring managers. It will certainly make it easier to build that connection which could one day become an offer!

As mentioned earlier in this chapter, GitHub isn't the place to put your resume. No, that's LinkedIn. But that isn't enough in today's day and age. You must also consider building your presence on social media, something we'll cover in more detail in the next section.

HAVING STARTED your LinkedIn profile and opened a GitHub account, it's time to start sharing bits of your coding skills on social media.

Now what, exactly, are we talking about here?

Simple. There is an old adage that reads, "If you're not paying for it, you're not the customer; you're the product being sold."[1] In essence, if you're simply scrolling through X, Facebook, Instagram, or YouTube and consuming en masse, then you're not really accomplishing anything. Instead, you are the target of all of its influences.

As a pro coder you should strive to turn the tables and *produce*. That's right, create content on social media and let *others* consume what *you've* made.

And there is nothing simpler than commenting on a tutorial or assignment you just completed.

Whether you just completed a lesson from a class, a tutorial prescribed by your instructor, or watched a video on YouTube, open up a tab to LinkedIn and write a simple review.

Now watch out! This is ***Rule 1: Curate a Professional Persona***. Emphasis on *professional*.

This isn't the place to haul off and slam the source material to the ground. It's okay to give objective feedback. But write this as if a hiring manager had stumbled upon your post and was deciding whether or not to send you a direct message (DM) about an opening. And if that's not concrete enough, try to imagine being in the same room with the author and speaking face-to-face. How would you communicate your opinion in such a situation?

That's it!

With that in mind, I'm sure you can offer your own opinion, with professional decency, and it won't take more than fifteen to thirty minutes at the most.

A week later, do the same thing. Think about every piece of technology you worked with over the past week, focus on one thing that caught your attention, and write a simple post about it. Start doing that.

Because what you're doing is threefold:

•You're training your brain (and fingers) to not just write code in a tutorial, but to analyze it after the fact.

•You're curating content that a potential hiring manager, colleague, competitor, or classmate can look at and provide some impressions.

•You're learning how to produce more than code, a skill you will certainly need.

A big difference between a pro coder and a mere hobbyist is your ability to analyze the situation. Things break. Stuff falls apart. Systems go down. And we pro coders have to figure out what happened. A hobbyist can sign off and go watch a movie.

Training your brain to analyze code you wrote and "retell" it to someone else is one of the strongest tools at deepening your comprehension.

Now we just talked about creating new content on LinkedIn. This doesn't mean you have to forgo everything else. In fact, it's quite easy to take your LinkedIn article and transform it into an X thread!

Remember earlier how I mentioned that magical tool of "reused content?" At the risk of "reusing" that idea, I'm repeating it here. There is absolutely nothing wrong with retelling your experience on LinkedIn, X, Reddit, and other places. It's fine to start with just one (LinkedIn being my recommendation) and later expand to others.

> NOTE: If you're wondering about certain other platforms (*cough* Facebook or TikTok), then I'd say pass. Facebook has always been notorious for being the platform where the content you create is pretty much swallowed up and gone before the day is done. I have found little value in posting stuff there. And TikTok seems to be a bit off the chain, at least when it comes to tech-based content.

The more conducive platforms to sharing your opinions and reactions are LinkedIn, X, YouTube, and Medium. These are platforms that do the best job of letting you create content, getting it out to people that are interested, and all with little setup on your part.

You are free to standup a blog site, but there are inherent issues with that. I'm not talking about technical problems. The problem with a personal blog site is that almost no-one will read your content if that's your approach. At least, not without deliberate action that I won't go into here, since that's less about being a pro *coder* and more about being a pro *content creator*.

Now one of the things that can rise up in all of us is a certain type of fear that can have a crippling effect. Maybe you have already tasted it? We'll cover it in the next section.

ONE OF THE most insidious things we as developers have to deal with is known as "imposter syndrome." It's that feeling that you're simply not good enough to have earned the position you have. Or that you're not good enough to aim for a certain tentative position. Or that you aren't qualified to post your opinion about a given tool or framework on social media.

This feeling is exacerbated by several factors.

One of the biggest culprits is social media. It's quite common for other people to post their "good" days but not their "bad" days. You made a mistake today, but the ten other people you follow all shared their success story today. They all seem to have it together. Why don't you? That doesn't help and ends up magnifying your own mistakes.

Another tricky factor is how our industry keeps changing. The precise skills you need today aren't the same skills you would've needed a decade ago. And they probably won't be the same ten years from now. An HVAC contractor has to learn a certain set of skills, but once they have mastered that, their success really comes down to finding clients and booking work. They don't suffer from their practices and expertise being completely supplanted every five to ten years.

For you and I, the landscape is constantly changing. New toolkits are always coming online. Old ones fade into oblivion. Neither a CompSci degree nor certification from a bootcamp program guarantee your future.

Of *course,* it makes sense to have some subtle fear and trepidation.

And to be completely transparent, I have felt imposter syndrome myself. Many times.

I once worked for several months on a solution to enhance Spring HATEOAS, the project that help your build links to support Hypermedia As The Engine Of Application State (HATEOAS). When I dialed into a Zoom chat to demonstrate my proposal to the project lead, it didn't take five minutes to realize my proposal was junk. Not some soft reaction like, "Gee, this could use some cleanup." Instead it was more like, "I actually don't like this entire approach. It's complicated, hard to work with, and doesn't really capture the original purpose of the feature." Sometimes it's better to just start over!

If that doesn't shake you to the core, you're just not human.

All this to say, imposter syndrome is *real*. It's not something you can simply cast aside as pure imagination.

We are human. We aren't perfect.

So *how* can you ward off this struggle? I actually stumbled into a solution that has worked well for me. Maybe it can work for you?

When I was on the Spring team, I was working with my manager on organizing my tasks. I was struggling to coordinate between the different modules I was responsible for and ensuring the right amount of time was being afforded to each. My manager wanted to help me be effective. And in order to give him something more than a few bullet points, thrown together for a ten-minute weekly Zoom chat, I proposed creating a work diary that I would let him see.

I fashioned a Google Doc where at the top of the document, every day, I would write down what I did as bullet points. Nothing deep. Just a quick summary of what I had tackled. If I worked a ticket, I wrote it down. If I even peeked at a ticket, same thing. I also tracked meetings, discussions, and times that I requested feedback on pull requests.

Basically, I started tracking everything I did.

And that changed everything.

Because after a few months of doing this, one day, when I felt like I was spinning my wheels on a particular solution, I happened to sit there, staring at this document.

And I stopped thinking about my current crisis. The feeling of frustration that was welling up within me...vanished.

Instead, I just scrolled through this document. Right in front of me, in just a few seconds, I could see *everything* I had done in the past week. More scrolling, and there was a solid month of action.

Because of the repository-nature of ticket link URLs, I quickly spotted how my times my focus had been split between different modules. A very deep sense of satisfaction filled my core. Instead of that insidious feeling of being in over my head while being surrounded by teammates with incredible skill and resolve, I instead was flooded with a realization that I was making a difference. A *huge* difference!

I was no imposter. I was, in fact, a rock-solid software engineer!

And I've continued maintaining my journal to this day. Not because someone else needs it, but instead because *I* need it.

And so I suggest the same to you. You don't have to share it with your manager, or anyone. Simply keep it for yourself.

The important thing is to start today.

Because another wonderful thing comes from this.

At any point in time, should your manager or team leader come to you and ask, "what did you tackle in the past quarter?" Low and behold, you will have a material record of your actions!

And as valuable as having a record of your actions is, the most important thing is that you no longer will be relying on your memory to combat that feeling that you don't have what it takes for the job at hand.

Your own record can affirm your skills and accomplishments.

Now there is one other key thing we need to tackle when it comes to ***Rule 1: Curate a Professional Persona***, and it involves your teammates. Check out the next section to find out what is vital to being part of a strong team.

ONE OF THE most important things to curate as part of your professional persona is the ability to trust your teammates.

Now what in the world does that mean?

It means that you're willing to listen to their feedback. You're willing to accept their changes to your code. And it means you're willing to support them...even if you think the path is wrong.

Well that got real serious, real fast.

It's a special thing to be on a team where you can professionally respect your fellow developers. And I'm talking about the people you work side-by-side with, writing code.

I once worked for a manager who, at the time, was writing as much code as me. And he did it better. When he gave me feedback, I listened. Because he wasn't quoting dusty textbooks or parroting trendy stuff. In fact, he was a bit of a skeptic of "rules of thumb" and pop CompSci. Whatever feedback he had for me, he could expound upon with detail.

We all have differing skills. Your colleagues have their strong points, and so to *trust* someone means to genuinely listen to what they have to say.

I once heard of a comparison between *belief* and *trust* and it goes like this. There is this guy that had a high wire act in which he didn't just cross it...he would carry across another person in a wheelbarrow. *Belief* is when you believe that the man can perform his bit. *Trust* is when you're willing to be the one that sits in the wheelbarrow doing the routine!

Maybe you believe your colleagues can write code, but are you willing to trust them enough to apply *their* suggested change to *your* code?

One of the most valuable assets to accumulate in your career as a software developer is experience. We don't always know how things will go, but the more systems we have seen, built, watched fail, and fixed, the better we get at sensing what to do next.

Your team is counting on you to bring your unique set of experiences to the table. This means providing sincere (but not condescending) input to code reviews and design discussions. And this also means learning about the experiences your fellow teammates have.

Something I learned was that one of my colleagues had worked for over ten years with Hibernate (object relational mapper) technology. I didn't know this when I first met him.

This colleague had given a talk at the 2018 SpringOne conference titled "*Spring Data JPA from 0-100.*" It was an hour long, and I watched it in my hotel room in 2023, the night before our team got together for a week-long face-to-face gathering. His talk had a series of "What do you think Hibernate does here?" questions he would pose to the audience. With every question he posed, I had no clue! By the end of his presentation, I felt a rising sense of fear.

Having taken the mantle of lead developer for Spring Data JPA in early 2022, not knowing the answer to a single one of his challenges didn't sound good!

And it shook me to the core.

Instead of seeing this fellow developer as someone that

came from a non-Spring background, I realized his experience with Hibernate was an asset for our team, even if he wasn't the lead developer.

I soon began trusting him more and more to review my proposed changes to Spring Data JPA. And he could read any ticket and quickly divine whether or not it was an issue involving our code or a bug in Hibernate itself. Considering the time zone he lived in was seven hours ahead of mine, he helped whittle down the volume of new tickets I had to actually address. (If you're wondering why he wasn't the project lead, that's because we needed someone to lead Spring Data JDBC, and his experience with ORMs was vital for the task!)

My ability to trust in my colleague actually took away much strain and made me realize how thankful I was to have him on our team. While it might have been tempting to see this person as competition, I instead came to value his input to my work.

Can you say the same for your teammates?

If not, what is getting in the way of building such a bridge? If you're struggling to connect as a team, perhaps you should consider carrying out some team building exercises. During that 2023 face-to-face team meeting that I mentioned earlier, we actually conducted several exercises to better break down walls and instead build stronger relationships.

During one of those exercises, we took the time to answer a series questions including:

- How did you get into Spring?
- What was the scariest thing you ever encountered?
- What is your favorite thing to do?
- What is the most amazing thing you've ever done?

Not every question was tech-oriented. Nor does it have to

be. What's important is sharing more about yourself with each other, and learning more about them. The more human every one of your teammates is, the easier it becomes to trust them.

We all make mistakes. But with a team in which you've built a deeper connection and are willing to listen and respond, it's possible to overcome mistakes and grow.

And growth is an important thing we need in order to curate our professional persona.

And with that in place, the next thing you'll need to become a successful pro coder is the right environment, which we'll cover in ***Rule 2: Build a Workspace***.

---

1. Andrew Lewis, https://twitter.com/andlewis/status/24380177712

RULE 2: BUILD A WORKSPACE_

Being a pro coder means taking this whole endeavor seriously. After all, this isn't some hobby. It's your livelihood! And it's even more than that. If you want to be a pro coder, I presume that you really want to ultimately make a difference in the world.

And so one of the most important places to focus on is your work environment.

But why?

Tell me if the following sounds familiar. Are you dabbling with code while sitting on the couch? Are you chugging through tutorials camped out at the dining table? Perhaps you're watching coding videos while sitting in bed?

All of these things show a common factor...it's a hobby. Hobbyists and amateur enthusiasts tinker wherever.

To become a pro coder and succeed, you must embrace ***Rule 2: Build a Workspace***.

That's because pro coders, in order to focus on building sustainable code, will carve out a place to dedicate to their career. Which of these do *you* want to be?

In this chapter we're going to dig into various tactics you can adopt to make being a pro coder a reality.

Two things that are tightly coupled to each other include carving out a dedicated space to write code and *also* carving out the time to do it.

Maybe you're in the middle of a four-year university program or perhaps a twelve-week bootcamp class? Or maybe you finished high school and entered the workforce for some job, hankering to get out there, but are now seeking a way out? Either way, it's vital that you pick a dedicated place to focus on writing code.

Whether you're a full-time coder or not, you need dedicated space. And here are some tips to consider.

If you live in a tiny apartment, think about picking a particular corner and calling it The Coding Zone. You can probably find a tiny desk either at some thrift store or you can spend the big bucks and visit Walmart!

You also need some sense of a chair to sit on. Now be careful! I've seen chairs that cost over $1000. You can really spend the big bucks simply sitting. You don't have to start there. A folding chair or one of the chairs that comes with your dinette set will work on Day 1. Of course, it doesn't hurt to save up and get something a *little* nicer.

My favorite tactic has been to visit a store that specializes in office furniture and sit in *every* desk chair that they have. I want something that's solid, comfortable, and has good back support. Picking an afternoon to go chair shopping isn't a bad idea.

Did I mention that anything you purchase out-of-pocket to build your coding space is *probably* tax deductible? Consult your favorite accountant or tax attorney to discover your options!

You need to build a workspace as your battle station, if you will. But you don't need to go into debt to do it. Instead, start small and find things that are inexpensive but functional. And then move things around in your living space to make it work.

And no, you don't have to call it The Coding Zone. (But you can if you want!)

This whole section may have sounded very niche if this book were released before 2020 because the number of remote positions was just a slim fraction of all positions available. But as of 2022, according to StackOverflow.com's survey results, less than 15% of software developers are required to report in-person full time. Almost 43% work fully remote while 42% work in a hybrid setup.[1]

The point is, when you pick a space whose sole purpose is to build apps and obliterate bugs, it actually helps you focus your thoughts on the task at hand. If you happen to live in an apartment or house that has multiple bedrooms, you can turn one of those into your dedicated office.

Which brings me to the next point...carving out the *time* to do it.

If you've been hired to write code, your company will have specified the hours you're expected to be working.

But if you're still trying to build that bridge into a new career of software development, *you* have to carve out the time to do it.

Best-selling author Jerry Jenkins once said that when he was launching his author career, he deliberately picked hours he could dedicate to writing or it never would have worked. He had a day job, so that wasn't an option. In the afternoon/evening, he wanted to focus on his wife and kids. So he picked 9:00 PM to midnight as the time no one would disturb him. And he started writing.

Carving out time dedicated to building your career requires thought and action. It won't just happen. While it may be incredibly tempting to binge-watch the latest season of the Foundation series on Apple TV+, go out with your friends to the movies, or a thousand other things, there is an underlying question.

What do *you* want to do?

Becoming a pro coder requires commitment on your end.

But even if you set aside the space and the time, there is yet more to becoming a pro. You need the tools.

ASSUMING you've set up your workspace, the next thing is to setup your work machine. Writing code is for naught if you don't have all the tools needed.

In this day and age, just about any compiler is free of charge. For half my career, I worked with Java, so to anyone that will listen, I always recommend *sdkman* as the tool to install. It easily lets you install and manage multiple versions of the Java compiler, known as Java Development Kit (JDK).

If you are using another stack, like Node.js, Python, C++, or Rust, then don't worry about *sdkman*. Find the right compiler needed to carry out your assignment or do your job.

This also demands that you have a machine for coding. I have used Macs since 2010 when I joined VMware, and I never looked back. I bought my own MacBook Pro in 2023 so I could have a machine that was completely under my control. I even convinced my tech-challenged wife to take a shot on a MacBook Air a few years ago, and today she'll tell you she can't work without it!

Nevertheless, if you prefer either a Windows or Linux machine, then get what suits you.

If you have secured a paying position on a team, your company should be providing you a machine. There's a chance they're going to ship you one, no questions asked.

But it's possible that they may give you a choice, or if there's any chance, you could simply ask, "What options do you have on picking my work machine?" When I joined Cockroach Labs in 2024, they gave a list of options, and I picked (you guessed it) another MacBook Pro. I have also had many colleagues with machines loaded with Linux. And a small sliver using Windows.

Whatever machine you choose, learn it well. This will be your livelihood. And that means take care of it and respect it.

Another thing I like to recommend is getting a rock-solid Integrated Development Environment (IDE). It's an understatement to call it an editor with the compiler baked in. That's what IDEs were thirty years ago.

Today, they are an incredible performance multiplier. That's because they have contextual awareness of your code and make it possible to navigate from the variable you're looking at to the line of code where it was populated to the line where it was declared or passed in as an argument.

An IDE can easily look back and see what invoked a given function. They can also spot mistakes and highlight them. And as you make corrections, the highlighting vanishes, making it almost a game! I've even seen IDEs that will parse SQL database queries embedded inside strings and point out errors.

I personally recommend IntelliJ IDEA[2] by JetBrains. The Community Edition is free. Of course, there are other popular ones for Java developers including Eclipse. And there is also VS Code by Microsoft. These are also free.

If you're using a different stack, spend some time digging

into what is the best IDE in that community. If you happen to be thinking of JavaScript, JetBrains also has WebStorm; if you're going down the path of Python, checkout PyCharm.

Of course, not everything has to be IDE-based. Sometimes you need a plain old text editor that works with you not against you, and for that I strongly suggest Sublime Text[3].

For a plain old text editor, it can sometimes fool you into thinking it's actually an IDE! It has partial text-completion, some amount of symbol navigation, and shifts modes based on what type of file you're looking at. In fact, I know a handful of people that use Sublime Text *as* their IDE of choice.

In this day and age, you also can't avoid the value of having Docker Desktop. Docker is one of the best step-ups in power that has emerged in the past ten years. Basically, Docker works on the paradigm of putting *everything* inside a "container." Every container has the same handles. Basically, one set of commands to start, stop, and interact with this "container" wrapped around an application. What app is inside the container? Anything you want!

Every database vendor out there is publishing containers with their database inside. With a single command, you can pull down a CockroachDB database server. Another command, and it's up and running. No more combing through gobs of manuals simply to install a database.

And it doesn't stop there.

Every form of middleware from RabbitMQ to Apache Kafka are available at your fingertips. You can find the exact version you need built on the operating system you want. So go ahead and install Docker Desktop[4] and setup your own personal account for free! Docker Desktop is also useful for test purposes, which we'll go into further detail in ***Rule 6: Test What You Build***.

You'll also need a browser. What developer in this day and

age doesn't have one? The machine you have will come with one. If it's a Mac, the built-in browser is Safari. Windows has Edge. Ubuntu Linux has Firefox. You can certainly continue with these, but my personal favorite right now is the Brave browser due to its default secure nature. But many browsers can also be hardened through plugins.

Another tool that I have for my workspace are AirPods Pro.

Maybe that sounds weird?

You were probably expecting another software app. Hear me out. AirPods have been game-changing. For years, I've had wired earbuds that come with my smartphone. And for years, they have gathered dust in my nightstand. Because they're too much of a hassle to get out and use.

AirPods in their recharging case easily fit in my pocket. I tote them around everyday. Whether I'm sitting at my desk, in a coffee shop, or in the car waiting to pick up the kids, I can play my favorite coding playlist. I can easily take a break and watch a YouTube video without worrying about disturbing those around me.

They are so easy to use that I actually *use* them. Check them out[5] if you're interested. (Yes, that's an affiliate link. If you buy a set, I'll get a small commission without it costing you anything extra.)

With these things set up, you can find other tools as they suit you. For example, on my Mac I've had Alfred for several years. It is a slew of productivity tools accessible with a single hotkey, including the ability to find any document on your machine assuming you know some fragment of the file's name.

Now with your workspace and machine setup for coding, it's time for a very important thing to be built into your work environment.

Coding is fun. It's cool. A total geek-out, I know!

But there's something it can also do...drag you in. Suck you into its grasp.

And if you're doing it on the clock, it can be easy to burn yourself out. Even if you don't ever suffer from an official "burn-out" where you lost your ability to code, you can still overdo it.

That's why it's more important than ever to draw out a clear separation between work and personal time. And even when you're "on the clock," it's okay to take breaks.

What constitutes a break?

Simply stop coding and go watch a fifteen minute YouTube video. Feel free to visit https://youtube.com/@ProCoderIO if you are shopping around for a channel! You can also read a chapter of that book you've considered. Or, you can spend a little time picking things up in the living room. Perhaps go outside and enjoy some fresh air.

And here's a biggie...I give you permission to pause and watch one episode of a TV show you've been interested in. You obviously can't write code for fifteen minutes and then chase it down with a forty-five minute show each and every day. But it's okay after two hours of coding to stop and watch that Star Trek episode you've been hankering after.

And if you let yourself take these breaks, you'll be amazed at how it refreshes your mind.

There is also something more official you can also consider: **The Pomodoro® Technique**. Yes, it's trademarked. I didn't make that up! The Pomodoro Technique in its simplest nature is:

1. Choose a task.
2. Set a 25-minute timer.
3. Work until the timer goes off.

4. Take a short break (5 minutes).
5. Every 4 Pomodoros, take a longer break.

This technique attacks two things: focusing on things we agree need to be done *with the understanding* that we can do *other* stuff soon. Sometimes you want to check X. Sometimes you want to work on that other *thing*. Sometimes you want to order a pizza or check your personal email.

All of these distracting things can pop-up at any time. And instead of succumbing to the impulse or beating yourself up for not "working hard enough," you can instead turn time from being a sneaky monster into an ally.

You can buy official timers or simply search Amazon for some generic timer.

Or, you can use a music playlist (check out https://procoder.io/coding-playlist if you want an 80s synth-heavy instrumental mix!) When the playlist finishes, take that break.

And also noted by fellow coding aficionado Russ Miles, it's vital to force yourself to take that break.

Another option is to go outside and take a walk—take your dog if you must! A notable colleague of mine, I discovered, would typically take forty-five to sixty-minute walks daily. In fact, we were at a team meeting when I learned that he was going out everyday, either before we gathered or in the evening, and simply walking around the streets of downtown Boston. He had made it his routine and didn't sacrifice it simply because he was away from home.

Some of my friends like to pack their running shoes whenever they travel to a conference. They can't wait to do their daily run in a new city.

To get there you have to start somewhere, so how about now? Perhaps consider putting down this book, getting up, and going outside right now to take a walk. Walk to the end of the street or to the light at the end of the block.

Don't bite off too much. Go ahead and take a shot. I'll wait.

Back? How was it? Now try this, before tomorrow "starts," do the exact same thing again. If you have a roommate, spouse, or partner, let them know you are going for a walk and will be back.

I want you to commit to doing it today and tomorrow. And then when tomorrow rolls around, and you carry out that promise to yourself, commit to doing it again.

Dr. Maxwell Maltz was a surgeon that once published how it usually takes people at least 21 days to adjust to a new routine. While your mileage may vary, the benefits over the life of your career to simply take a daily walk can be long-lasting. I encourage you to invest in yourself and give yourself time and grace to go for it.

I've tried it myself, fallen off the wagon, resumed, and slipped out of practice again. So let's take up this challenge together, okay?

All these various ways of taking breaks have an amazing side effect. When you return to your desk to continue coding, you may have an epiphany! Sometimes, you realize what you need to do that stumped you earlier. Or you may discover you were going down the wrong path and can change course.

That's because your brain often continues working on the problem you started.

And one other thing...the most *important* break you can take is at the end of your work day. That's right. Hang it up. Go home. Stop working.

I understand you can feel passionate. There are times I've worked late. I'm absolutely "in the zone" and want to land a new idea.

But that should be the exception and *not* the norm.

To be honest, I have several friends that have been in this industry for a long time, and have recognized that getting in four hours of solid coding in a given work day is about the max. And I agree.

Some days, I can't make the code "click", but I can certainly disposition lots of tickets!

It's 100% okay if 4:00 PM rolls around, things aren't clicking, and you decide to hang it up early that day.

Logoff, grab your keys and phone, and go home (or leave your home office).

Top-notch developers do *not* write code continuously, for eight straight hours. It simply doesn't happen.

No amount of "we need this yesterday" coming from management can make you more performant.

So what do you do if that's literally what your manager is telling you?

Ask your manager, "what do you need first?" Make them decide what's most important and signal that you will give it top priority. That's fine. This is what management should be doing, prioritizing tasks.

But it's up to *you* to tell them what you can and cannot do.

In fact, I remember a proposal years ago at a previous company. The VP, prepping to talk to the customer, was chatting with our beloved Chief Software Engineer. The VP asked, "what can we do and not do?"

This confused me, a fledgling developer early in my career. I thought management *told us* what to do!

The VP went on to say, "I need you to tell them. They won't believe me."

This was a powerful realization for me, and it's this: customers are more likely to believe engineers.

Managers are always a step away, if not more. The engineers that are slugging out the code are the ones "in the know."

Engineers are the ones that can describe what's possible and what's not possible. Engineers are the ones that can estimate what can be built and when it can be built.

Do your part and indicate what you can do. Don't be pushed into saying something you don't believe in.

And so, it really is up to you to first be honest with yourself, so you can then be honest with your manager about what you can accomplish.

If you think that spending the next eight weeks working eighty hours/week is the solution, then you're setting yourself up for a huge failure. It's okay to tell your manager their request is going to take eight, twelve, or sixteen weeks. Whatever it takes.

In fact, this can be a signal that the scope is too big and should be chopped up into smaller pieces.

And none of this...NONE of this has to be done without those daily walks and other breaks. Your daily walk combined with breaks will actually refresh you. It can actually empower you to work more efficiently and have a stronger sense of what you *can* accomplish in a given time frame.

Don't be afraid to stand your ground. You owe it to yourself.

Now that you've built a work environment including your machine and also a plan for taking breaks, it's important to learn the next rule, ***Rule 3: Sell Your Talent and Experience***.

---

1. https://survey.stackoverflow.co/2022/#work-employment
2. https://procoder.io/intellij-idea
3. https://procoder.io/sublime-text
4. https://procoder.io/docker
5. https://procoder.io/airpods-pro

RULE 3: SELL YOUR TALENT AND EXPERIENCE_

We're talking about being a pro coder right? Well, at the heart of all this is the word "pro," short for "professional." We aren't talking about a hobby. Instead, we are figuring out tactics to make this a fully-paying career.

I'll just presume you have a passion for coding. But the thing to remember is that you *deserve* to get paid for that. Let me repeat this key point, in case it flew past. You *deserve* to be paid properly for embracing a field with some of the most abstract, dynamic, ever-shifting concepts humanity has ever seen.

And so with ***Rule 3***, you will learn to ***Sell Your Talent and Experience***.

Before you can get that paycheck for building complex systems with code, you first need to earn the position. And that means having to enter the foray of job interviews. A topic *no one* enjoys.

To be honest, the whole prospect of doing interviews is

quite nerve-racking. At my alma mater, the registrar's office had a program to coach soon-to-graduate students on interview techniques. I did a practice interview.

On camera.

And then I watched the whole thing.

Aye-aye-aye. It was *horrible.*

I watched myself fiddling with my class ring. I kept shifting my feet around, propping one leg up on the other. My whole body could not seem to *stop.*

Armed with this information, I traveled to my first real, on-site interview. And while I was subjected to four different interviews, each with a different person, I sat and consciously kept my feet on the ground. No shifting them back and forth. I also put my hands on my knees and didn't move them.

I held all the energy inside and focused on answering questions.

And a few months later, I got an offer! After a couple more interviews with other shops, I ended up accepting the first offer.

I'm sharing this story, not because it's everything you need to do to land a position. Hardly. The idea is that sometimes you need to go out of your way to accomplish a goal. Be prepared to take whatever feedback you can find, and apply it.

Some additional sneaky advice I'm about to share that no one seems to talk about in blogs or videos is what teams are *actually* looking for. You see, just a few years after starting that position, I received a promotion.

And one thing my manager told me was that I was expected to partake in the interview process. Not wanting to get dinged on my own annual performance review, I embraced this call to action and ended up interviewing dozens of prospects.

At that company, a candidate was subjected to four

different interviews. Then we'd all meet at the end of the day, compare notes, talk about issues, and then make a recommendation to the department's engineering manager.

And I'm here to share *exactly* the criteria we used to choose whether someone was qualified or not.

The number one thing we looked for...were examples. But of what?

Did the candidate have examples of using technology, of solving problems, of working with others? The methodology we used at the time was known as looking for **STAR**s (**S**ituation, **T**ask, **A**ction, **R**esults). Basically:

- **S**ituation — What is a situation that the candidate found themselves in?
- **T**ask — What were they required to achieve?
- **A**ction — What steps did they take?
- **R**esults — What was the outcome?

And to be transparent, we told every candidate that we interviewed, that this is what we were looking for. Instead of grilling someone on whether or not they knew how to build a linked list, we instead wanted something bigger and more emblematic of solving problems.

Anyone can look up the details of method calls in a reference manual. Or can find the algorithm to the bubble sort algorithm.

The real question is, do they know *what* to look up and *why* they might need it?

We weren't hung up on whether or not they used the same technology as our current contract. New hires can be taught the technical stack relatively quickly. The real question is...if taught the specifics, do they have the ability to effect change that will help out the team?

Being familiar with the interviewing process later helped me secure a position on the Spring team back in 2010, despite the fact that writing Java code was not my primary responsibility at the time. It may sound strange to work your way onto a team so heavily steeped in Java-based technology without ever having been responsible for a J2EE application (the forerunner to today's Jakarta EE)!

It's true that I had worked for a decade using a proprietary non-Java technology, responsible for a network management app used by our Ops Center!

To dial things up a bit, at least three of the phone interviews I was invited to involved members of the Spring team I'd likely be working with. These people were senior-grade Java experts and would quickly spot substandard quality!

So I leaned into my own awareness of the value of solving real problems.

I was upfront and honest that while I had not worked directly on Java nor with any web technologies, I had extensive experience solving midnight failures. I had been able to manage seven people working directly for me, and had resolved countless tickets with our 24x7x365 operational team.

I built a compelling case that made me irresistible!

The next time you take a break, perhaps this evening, sit down either with a pad of paper or with a blank Google Doc, and start listing some hefty scenarios *you* have had to solve. Try to think of the hardest problem you have had to tackle during your university program, or perhaps the trickiest tutorial you decided to do after-hours and how you worked your way toward a solution.

If you have already entered the professional coding workforce, think about the toughest problem you've solved at work and write it down. Capture all the details you can think of.

And over time, update this document. It's why I recommend using something like Google Docs. As you solve bigger and bigger problems, let those instances move to the top of your list. Hone your list of accomplishments.

You never know when you'll need to tap them for an interview!

But that's not the only thing you need to know to navigate the interviewing waters. What if you succeed and get an offer? What then?

If you don't get an offer, that's the end of the road for that shop. At least for today. If you are given a chance, you can politely ask for any feedback on why you didn't qualify. Swallow your pride and listen. But don't try to beg your way in. That doesn't work.

The real gulper is what to do if you *do* get an offer.

If a team is interested in hiring you, they will extend you an offer letter that should include things like:

- **Salary** - your base pay, could be every two weeks, twice a month, monthly, or something else
- **Start date** - when they expect you to start working for them
- **Bonus** (optional) - some shops offer additional pay, often paid either annually or semi-annually, and often based on performance
- **Stock options** (also known as equity) - usually with lots of terms and conditions as to when they're available, how much they're worth, and sometimes with limits on when you can exercise or sell them
- **Benefits** - healthcare, PTO (paid time off), sick time, paternity/maternity leave, etc.

What's important to grok is that it's not an all-or-nothing situation. You have entered the **negotiation phase**.

It's important to understand that the team that wants to hire you is also trying to meet or come in *under budget*. And so they won't necessarily offer you the best deal (for you). That's why you can always ask for more money, more stock options, and other things as well.

When I got my initial offer from VMware, they had given me a decent salary offer and a certain amount of equity. But I had already been working for several years at my current position while providing on-call support. In my current position, I had received a 15% bonus payment for at least five years. I wanted a new salary that incorporated that *plus* a bump.

There are other factors like cost of living for the region you're in. San Francisco and New York City have discernibly higher costs than Florida where I was located at the time.

When you receive that offer letter, which is usually an email with an attachment, you **don't have to respond** to the offer letter instantly.

Let me repeat that a bit more clearly: don't respond *immediately* to your offer letter.

Instead, take some time that evening to sit down and make a list of all your current expenses. If you're looking at getting an apartment, try to estimate that. Work out all the money you need to cover things. Don't forget to include a line item for saving. 15% of your take home pay is a great place to start.

Then ask yourself a key question: does their offer even cover that?

A valuable resource to check out (ideally *before* you engage in interviewing!) is https://levels.fyi. There are actual consultants you can hire to help negotiate your offer. One service I retained in 2024 to help negotiate my latest compensation package is Team Rora. They provide coaching to help increase your offer, and only cost a $500 retainer fee + 10% of the difference between initial and final offer x 2 years.

Considering we're talking about something that sets the baseline for the next twenty five years, that's not a bad deal. I can certainly testify that there is a lot of emotion and butterflies-in-your-stomach feelings during the interview and negotiation process. The recruiter you interact with does this every day. You don't. That's a keen advantage for them. Having someone in your court to guide and coach without being subjected to emotion is of high value!

levels.fyi also makes it possible to search for the position you're considering in the area you're aiming for, and at the level they are offering to hire you at, and get some idea what others are making. It can either validate the offer they've extended, or show them to be way under market value.

Armed with all this information, including both your needs as well as evidence of market data, you can write back to the person making the offer something like this:

Dear <Person who extended the offer>:

I appreciate your offer. I'm excited at the prospect of working for Acme Inc. However, market research shows that Level 1 Software Engineers in the Pandora metropolitan area are averaging at least $XX,XXX. If you can increase the offer to $YY,YYY, I think we can call this a deal.

Sincerely,

A. Wolf

If I wasn't clear, *everyone* negotiates. *Everyone.*

Don't panic thinking they'll rescind the offer. They are actually expecting a counter offer and probably already have a response set up.

Maintain a professional tone, and send your counter offer.

But what if I don't go high enough?

This question is always in the back of my mind.

In truth, the ideal situation is that you make a counter offer *above* what you *really* want, and their response lands on your target amount. But it's possible that your over-the-top offer... may simply be accepted!

I'm not sure if I can get $95,000, but I'll take a shot asking for $110,000....What? They accepted $110K???

Heh.

In business when you shoot high and they accept immediately, it's a sign that you bid too low. A friend of mine didn't want to take on a particular contract being thrown at him. He bumped up his usual rates and demanded a business class flight to kill the deal.

They accepted in hours.

It's no secret since we discussed this during one of his live streams. We both chuckled, because *we both* knew he had countered too low.

If you're still too white-knuckled to consider negotiating at all, then consider this. There are actual studies showing that engineers don't negotiate enough. Yep. Academics are analyzing this, and we engineers aren't business savvy enough to play the game of salary negotiation.

Do it.

It's how things work. No one will be offended. No one is withdrawing an offer because you pushed back.

By the way, salary isn't the only thing you can negotiate over.

I mentioned earlier that I was also offered stock options. In a couple rounds of that offer's negotiation, the hiring manager kept upping the stock options. A nice perk, but I knew that stock options were a finite item and wouldn't last forever.

In one of my responses to the recruiter from VMware, I expressed that stock options wouldn't pay my mortgage. I kicked up my salary request to equal <old salary> + <on-call pay> + <an additional 15%>, and they accepted. (Too low?? Maybe!)

Don't forget to ask about the benefits. Healthcare, savings options, paid time off (PTO), the works. That type of stuff is usually baked in and not very flexible, unless you're aiming to be a Director (or higher). But anything with money linked to it is usually up for grabs.

Also, is there something you are currently involved in that you want to keep? For example, are you the leader of some open source project you do on the side? If you're not careful, the company could subsume your role on those things. If the project is wholly built by you, they could technically lay claim to it when you sign the **Non Compete Agreement** most companies will require you to sign.

You can include in your agreement something like:

> *Project XYZ will remain a separate project not under the purview of Acme Inc., and that efforts spent during non-work hours will not be considered as competing with Acme Inc. and its interests.*

Unless they're hiring you to work on that project, make it clear that you're not moonlighting and expending effort for that project on-the-clock.

Have you created a YouTube channel? Perhaps gotten it monetized? You may want to discuss that as well, ensuring you retain full rights.

Ultimately, the goal is to get on the team and to be fairly

compensated for your efforts. So keep reminding yourself that soon enough, this tension of haggling over your pay will pass and you can get on with the cool stuff.

Just don't shortchange yourself. Always be willing to aim for the stars and ideally land on the moon.

Now WE'VE TALKED QUITE a bit about strategies on navigating the dreaded interview process. Hopefully, you won't have to visit that *too* often.

However there is something you may not have considered that warrants your attention *every year*.

Raises and promotions.

There is a simple concept present in every tech firm I've seen and it's called the **salary band**. A salary band is simply a range of dollars that a company pays a certain level of developer. For example, if Widgets XYZ were paying their Level 2 Software Engineers anywhere from $80,000 (lowest paid person with that title) up to $92,000 (highest paid person of the same title), that range would be that title's **salary band**.

Now the following aspects of the salary band I'm going to share are *vital* for you to understand, so lean in a little closer.

If you're *below* the midpoint of your salary band, there is an upward pressure to give you a raise. Why?

- If the company does *not* give you a raise this year, the risk is that you'll walk out the door into the arms of one of their competitors.
- Because you are *below* the midpoint, giving you a raise has little impact on the salary band.

Ask your manager if you are permitted to see your salary

band. While companies don't typically share their salary bands *publicly,* they may allow employees see them. At my old company this was shared internally with all employees. There is a fair chance your current company will share it with you.

In the example we just had, a Level 2 Software Engineer salary band ranging from $80,000 to $92,000, the *midpoint* of that salary band would be $86,000. If you were making *below* $86,000 then there is more pressure to give you a raise than if you were *above.*

Figure out which salary band you are in and how far you are from the midpoint. If your manager conducts any sort of 1-on-1 meeting and asks if you're happy with your current pay, tell her "no." Maybe that sounds rude? Or perhaps downright... wrong to say?

Just to let you know, it's *always* fair to signal that you're struggling with bills, rent is going up, you are having to pay for kids, or you're dealing with inflation. Don't lie, but don't hold back expressing that you would welcome a raise.

Rest assured everyone else in your department is giving the same signal to their manager when the opportunity presents. Some even if the opportunity *doesn't* present.

Don't get left behind.

Don't be rude. Don't be petulant. And don't give off any sort of entitlement.

Instead, simply indicate the needs like I just described, and then feel free to follow that up with a more conciliatory expression like, "I want to ensure I'm providing value to the company and serving their needs. A salary increase would certainly help me remain focused on delivering!"

Get it?

Now I mentioned that there are several important things to understand about the salary band. And here's the next one.

Remember how being *below* the midpoint puts an upward

pressure on giving you a raise in order to encourage you to stay and not force them to go through the hiring process again?

Guess what...if you are *above* the midpoint, the pressure is reversed. When you're above the midpoint, every raise they extend to you has a much more noticeable effect on the upper bound of that salary band. They *do not* want that!

Let me repeat. Your company does *not* want the upper end of any salary band to go up. Because then they'd had to give *everyone* a raise. No, they'd rather hold you back and hold that salary band where it is.

What can you do about it?

There's gold at the end of this rainbow, because the next step for someone moving on up the salary band is...*promotion*!

Yep.

When you cross the midpoint, the pressure begins to mount to promote you. Why, exactly?

Because if they promote you, it moves you into another salary band. And being at the bottom of the new salary band means they don't have to worry so much about handing you a raise.

I know this sounds kind of weird, but this is how it works.

So the question you need to start asking your manager in your very next 1-on-1 meeting, after learning you are above the midpoint, is to ask, "What exactly do I need to do to qualify for a promotion?"

If you quietly, patiently sit by and wait for your manager to bring up the subject on their own compared to what I just suggested, or if you wait for the department manager to graciously extend you a promotion with no initiative from you, you are stunting the growth of your career.

Talking to your immediate manager about what you need to do is the right way to get yourself lined up with what upper management is expecting. You see, any department will have a limit on the number of promotions they can hand out in a given year. By finding out what's expected, you won't be floundering, forever waiting. Instead, you can focus on what's really needed up above and begin to present evidence that you're the one they should bump up.

Think about this, if you're not working off the same list as *them*, you'll not even make the list of candidates they are considering. You may actually be surprised at what you're expected to do.

Which is why you need to build that rapport with your manager.

One of the first things revealed to me when I got my first promotion early in my career was their expectation of me to participate in interviews of new hires.

I was dumbfounded.

Hired to write code, I naively thought that's all there was to it.

Once I realized what was needed, I carried out as many interviews as possible. (And today, *you* are reaping the benefits!)

I just mentioned how the pressures work regarding pay raises and promotions. Does that mean you are guaranteed either in a given year based upon your position relative to the midpoint? Hardly. The overriding factor is the company's profits for the year. In essence, was it a good year? Is there enough money for the company to dispense merit increases, bonuses, or promotions?

Another factor is that they also tend to cap the number of people in every salary band. There may be 20% fewer software engineers in the Level 3 band than in the Level 2 band. And as you go higher and higher, there are more requirements than simply your pay. I once worked at a company whose top level was Level 6. I had never heard of it despite years of working at that company because there were only 2-3 people that held that level in a company with over 3000 engineers.

When you're talking about the top salary bands at a given company, the justification to promote someone into that band is far removed from the simple pay rates. At that point there are bigger factors like how much business does this person bring in, what about customer rapport, what has their entire track record been like, and other questions.

This is a bit esoteric, because if you're reading this book, then you probably aren't vying for such a title. So let's bring this back to talking about moving from Level 1 to Level 2.

The name of the game is to work with your manager to learn what you need to be doing to either qualify for a solid raise or to go the extra mile and get promoted. And the salary band with its midpoint is the secret decoder ring to see where you need to focus your efforts.

Now let's step back and consider a Real World™ example. This is a situation I can testify to, because it involves someone that once worked for me.

I had hired a crackerjack coder fresh out of college. He came in as a Level 1 Software Engineer. At that time at that company, being a manager of a group of about seven engineers, I knew what the salary bands were and I knew the salaries of everyone in my group. More importantly, I knew where they all stood relative to the midpoint.

At the time of this person's first annual review, I was able to walk into the manager-of-managers meeting and pitch this fine developer from the top of my "stack" of developers I was responsible for. I had a mountain of evidence that this person beat out everyone else in the department.

In fact, everyone in the room had already seen the results of his talent, and so he got a maximum raise. His first raise boosted him above the midpoint.

By the way, just about every company conducts this process, whether or not they call it "stacking." I think my old company even changed the name at some point, yet did the very same process up until I left. Different shops may have different names for it. They may even say they're NOT doing *that*. But they are.

All the mid-level managers are comparing their top candidates against each other and using that to decide how to split up the merit increase budget between everyone.

Well two years in, and there was no task I could throw at this guy in my department that he didn't execute spectacularly. I went to the following year's end-of-year review and played him again. No one debated.

I had also checked out our company's official promotion manual at the time, and it specifically said that promotion to Level 2 required a minimum of three years. Simply put, he wasn't qualified. Remember how I mentioned additional requirements as you move up the salary bands?

So we got him a top notice raise. Not as much as previously, because he was now above the midpoint. Nevertheless, he ended up close to the top of the band.

After his third year and with me giving him bigger and bigger responsibilities, I had already started crafting his promotion packet. There was so much evidence of performance, and the fact that he was at top of his salary band, that selling management on promoting him was no effort at all.

Having parted ways when I left to join the Spring team, I recently looked into this past colleague of mine. He had changed shops and had been made vice president!

To cover all the bases, I'll share that not once did he ask me about what he needed to do to get promoted. He was what I'll call a unicorn. A rarity. I certainly haven't seen anyone else with such a stellar performance.

Are you a unicorn who stands out boldly earning pay raises or promotions without bringing it up? Odds are you aren't.

Doesn't mean you aren't good at what you do. It just means you need to do the work I mentioned earlier in building that relationship with your manager to grow your career. Don't let yourself falsely think you're the best. Because even if you *are* the best...my advice is still sound.

Another nugget buried in this example is the fact that not only do you want to learn from your manager what is expected of you for your current title (and the next one!), but you want to build up the evidence that you're doing it!

You want to leave a wide swath of proof that your efforts are delivering top notch value. Don't cower in fear that you'll be laid off. Instead, *try* to make yourself *impossible* to not give a raise or promote!

WE'VE TALKED about the hiring process as well as the annual review process. And in both of these scenarios I mentioned the need to build up a mountain a of evidence of your value to your team and company.

In this section, we're going to discuss *exactly* how to build up those mountains.

It all starts with capturing your outputs.

Now I mentioned back in ***Rule 1 : Curate a Professional Persona*** the many values of maintaining a work log in a Google Doc. Something that you alone possess, but can tap into. Naturally the question arises, "what do I put in there?"

And one of things mentioned is links to tickets. The team you're on will most certainly have some form of issue management system, be that GitHub issues or some other ticketing system. No matter what system you have for managing tickets, there will certainly be a way to link to a given ticket you are involved with.

There are other tools that track work tasks for a given "sprint" or work cycle. Anything where you have a stated goal and you are meant to capture your activities as they are carried out. In fact, on certain level, these can all be referred to as "tickets."

So be sure to consult with your team leader or manager about what tickets *you* need to be working!

If this is early in your career, you may be assigned to only one project. And it probably has a single ticketing system. You may even have some form of weekly or bi-weekly meeting where all the tickets are assigned to teammates. That is where you can get your assigned list of tickets to work.

So...start working 'em!

Whether you are simply reading a ticket, doing research on the topic of a ticket, or actually carrying out code changes, all qualify as something to denote in your work log. For more detail on recommended practices regarding ticket management you'll have to wait until Rule 8: Manage Your Tickets. The key point is to start tracking your involvement with tickets as a way to grow your mountain.

Another way to add to your record is to look for ways that your team handles community outreach. The Spring team invested countless hours in supporting their international community. This included visiting Stack Overflow, the Q&A website, and answering relevant questions about its various modules.

Whether your team is using Stack Overflow or some other forum to serve its community, look for time to peruse that environment. Try to find questions that haven't been answered that you can help out with.

Don't forget to update your work log with links back to questions and/or forum posts!

Now I mentioned that the primary purpose in engaging with your team's community is to help build the evidence of you supporting your team. But there's a more subtle benefit here.

The more you practice answering questions about your tech stack, the better you will become with your tech stack. I once studied tae kwon do. My master instructor told me that the quickest way to learn something is to teach it others. He was right!

I met someone that had actually chosen to answer a question-a-day on Stack Overflow. After a couple years of doing that, they actually joined the Spring team. Their long track record of answering questions paved a way, illustrating their knowledge and skillset.

I also know others that are in the Spring team that have spent countless hours responding to community questions with answers. This is what open source is really meant to be about. An entire community of people helping each other. You don't have to know everything to help someone else. All you need to know is the answer to one question to help that person.

And let me stress that while I'm putting this strategy forth as a way to build your record of talent, it's not that transactional. The truth is, this is how you actually become good at something. And this is also how you establish yourself in a community. But contributing in a sincere fashion.

All I'm saying is to not hesitate to gather links in your own work log while you're doing it!

Now another arena you can do is one that has served me well—creating demo apps.

Anytime I get into a new toolkit or stack, I begin to build a simple app that uses it. I've done this for Spring modules, Java-

Script toolkits, and Python projects. Heck, I even wrote demo code when I read Practical Common Lisp!

And in this day and age, feel free to capture your various demo apps and throw them up on GitHub. Not because some recruiter is going to stroll by your demo apps, marvel at it, and ring you up with a job offer.

While that could technically happen, the odds are slim to none. Go back and reread ***Rule 1: Curate a Professional Persona*** if you don't remember why GitHub ain't your CV.

Instead, think of GitHub as the place to host your playground of demo apps.Now if you really want to put all your demo stuff off to the side and not clutter things up, then feel free to create your GitHub organization.

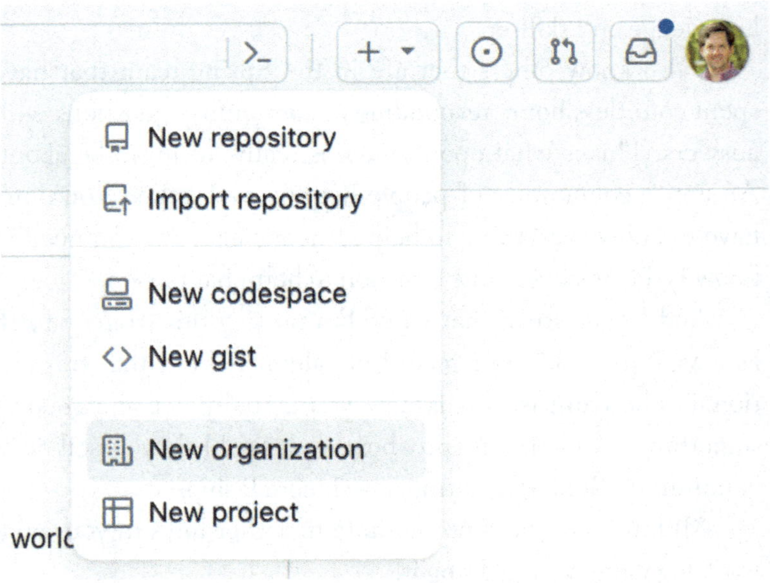

Back in the olden days, organizations were a paid-for item. But ever since Microsoft acquired GitHub, organizations are free. So feel free to use them!

It's important to share that these demo apps don't have to

be perfect. Far from it. The point is to tinker and fiddle with the code. That's how we learn.

In fact, back in the day I was asked to take over as the lead developer for Spring Web Services, the module to help interact with SOAP-based services. To get my feet wet, I built a multi-module project and I forced myself to learn how to build SOAP-based services and how to consume them.

SOAP is a relatively stable stack, so I didn't have to do *too* much. But every now and then, to get myself back in the vibe of Spring WS, I would dig up that demo project and walk through it. And it helped me get back in the game.

This is how demo apps can help you.

If you're super nervous, you can make your demo apps private in GitHub. (Thanks Microsoft!)

Now another area where you can further grow your knowledge base is writing blog articles. See if your team or company actually writes tech-based blog articles. It may be no big deal, or it may be a monumental deal.

In other words, the team you're on may write articles all the time. In that situation, I'd move fast to get into the game. Simply put, the more your team does something, then your lack of doing the same may appear to be a negative!

Now I didn't say write the article fast. What I'm saying is to not ponder the idea for a couple months. Instead, I'm suggesting you commit to writing an article right now and shift your focus to doing research and assembling the bits needed for the article. At the same time, you can ping either your manager, your task leader, or your website's editor (in case you have one) to see what it takes to publish something. Get the wheels turning!

However, you could be working for a shop that only publishes tech articles once in a blue moon. Or perhaps they only write marketing-styled blog posts, such as white papers and published studies. That sort of thing. If writing tech-based articles doesn't appear to be a thing, then there's an opportunity here.

To write a tech article where none has been written before will require a more extensive approval process. You'll probably have to pitch the idea to your team leader, take it up to a director, and possibly get a VP to buy into the idea.

You may also have to take a shot and actually spend the time writing the article in advance. Commit to doing the research to make this a top-notch, polished article. THEN attempt to take your article up the chain and see if you can get it approved.

You see, if you can blaze that trail and start your team or company publishing tech articles, that would be a remarkable feather in your cap since you'll have been the one to open that front.

I must warn you though. Blazing trails is harder. They require you pitching the idea to multiple people. And you'll probably receive a lot of pushback.

- "We've never done that before."
- "Our website isn't for content like that."
- "Isn't this marketing's job?"

These are just some of the questions you'll probably get. Be ready with an answer to each one. And some may come from within your own team!

But imagine the results if you are able to succeed! You could introduce a new wave of content. There are probably others on your team that have wanted to also write tech-oriented articles, but were unwilling to go through the headache. Or maybe they got shut down. If you can crack the door open, others may follow.

We've tackled many ways you can grow that mountain of evidence beyond simply writing code. We've seen ways to grow your knowledge while helping the community. And ways that you can become more than simply a code slinger.

Now check out the next rule, ***Rule 4: Know What To Do***, we'll go through the sorts of things you should be tackling on Day 1 of your new position, and what to do after that.

RULE 4: KNOW WHAT TO DO_

So you braved the interview process. After a few rounds with a hiring manager, you settled on your salary and benefits. And now you're *there*. So what now?

In ***Rule 4: Know What To Do***, we're going to cover some steps to take once you get that position, how to grow your presence on your team, and how to grow into your position and make yourself a powerful asset for your team.

THE MOST TENSE moment from your perspective will probably be your very first day. So take some notes. And while you may be filled with excitement tainted with a hint of panic, try to enjoy the process.

It's fair to assume that during your onboarding process, the IT department will grant you access to company email and possibly whatever instant messaging tool the team is using, be it Slack, Teams, HipChat, or whatever. Many of these tools can be accessed either through your browser or through an app. Download the tools as directed by IT and first try to access them. (Kind of hard to ask for help if you're not even hooked into email, ehh?)

I'll go ahead and point out that over the last decade, teams have gotten better at onboarding. Heck, it wasn't much of anything back in 2000. In 2010 when I joined VMware, it was a little smoother. My manager was asking for feedback so they could improve it.

When I joined Cockroach Labs, they grouped me into a cohort and we were all signed up for various meetings to meet the different parts of the company. I was also assigned a "roach-mate" with whom I'd meet once a week and guided me along the way. Fantastic.

Whatever your team has, pick it up and run with it!

Next, make a point of finding out who your direct manager is as well as your immediate teammates. **Write down their names.**

That's right. Use either your iPhone's Notes app or an old-fashioned sticky note. Anything will do.

Write down these people's names. You may never read that note again, but the very act of writing down this information will help commit it to memory. And knowing who your immediate teammates are is key.

Your direct manager may, in fact, be the person facilitating your onboarding process. Or your team leader may be doing that. It varies from company to company.

You also may get invited to a virtual room or channel pertinent to the team you've joined. As you see messages posted, take note of *who* has written those messages. You can probably click on the person's avatar and ideally see a picture of them as well.

While you are reading a flurry of messages, take it easy. Try to relax.

No one is expecting you to solve the world's problems on

your first day. Just try to drink it all in. Read every message. If there's a link to a blog post with comments, go read that article. If there is a link to a fragment of code, click on it and start reading the code.

Making changes to code begins with reading the code. If you're lucky, it could be something you're semi-familiar with. But that's not always the case.

Also, your new manager may have a set of actions for you to carry out including:

- Registration with various systems, like e-mail, benefits, payroll systems, and GitHub.
- Filling out some e-forms, like your home address, phone number, and other details.
- Other forms of "paperwork."
- You might be required to enroll in some online training classes that involve either orientation or some basic training all employees are obligated to take that could be related or unrelated to tech stuff.

Do it.

It's time to start showing you're a team player by carrying out your immediate assignments.

And above all, **listen**. You'll be getting fed a firehose of information. Don't sweat it, and simply listen and read. And if you're smiling with excitement at your new position, that's okay as well!

WE TALKED about what to do on your first day. But what are some key things to do in your first week?

Have you gotten in touch with the system administrator

and registered your GitHub id? Sophisticated teams may have already covered this in their onboarding procedures, but not everyone is this deft. Giving them your GitHub id will enable them to grant you access to the repositories you will be working on.

Here's a biggie: have they sent you a work machine yet?

If you're working remotely from home, perhaps not. Make sure the provisioning department has your mailing address so nothing on your end delays shipping that work machine. If you don't even know who to contact in the provisioning department, this is a great time to reach out to your manager.

Assuming you *do* have a work machine, have you cloned the immediate repository you are assigned to? Also check out the following list of things:

- Have you started reading the code? If you haven't been handed a specific task, you can start at the top and simply start reading the build files.
- Have you installed the right tools to build it? (Java, Node.js, Python, whatever)
- Inspected any of the continuous integration (CI) tools used to build it? (`Jenkinsfile`, `.circleci/*`, `.github/*`)

Now maybe that list I just gave you sounds like a LOT. Just do one thing at a time on each day. The idea is to start reaching in, running your proverbial fingers in the code and its supporting infrastructure. Start to taste it.

Again, the idea isn't to start building epic features, but to instead become familiar with how to build things.

Like I said, carve out plenty of time and start reading the code.

IMPORTANT: After the first week, you may want to start reading the backlog of tickets. Start getting familiar with the problems the users are running into. Once again, this isn't so that you're ready to solve them. Instead, start getting a feel for the pain points the community is hitting.

I HAVE OFTEN FELT that a month is the *minimum* amount of time for me to start getting the hang of a given project. This presumes I have been reading the code, building the artifacts (e.g., `./mvnw clean package` for a Maven-based system), and trying to nudge things here and there.

I mentioned in the previous section to start reading the code. This should include the reference docs, which many projects include inside their source code. If not, find out where they are and start reading those as well.

Because one of the first things you can *do* in your first month is to submit a pull request (PR) patching the ref docs. Fixing a typo, rewording a sentence. Anything along these lines. You may feel like an ant moving a pebble, but we all start from zero on a given project.

NOTE: Learn how to render the docs and inspect your handiwork. You can often figure this out if you read the CI files!

Other stuff you can be doing within the first month is spotting any sloppy code, old patterns (e.g., missing generic parameters, missing usage of the diamond operator, etc.), or unused import statements. Anything that simply "cleans up" the code without actually changing the code.

Maybe you're curious what, exactly, is clean code? Doesn't that sound a little subjective?

We'll cover that in more detail later on in ***Rule 5: Structure Matters***. For now, think of it as following the overall style that all the *other* code is following. I know that's a bit of a stretch, but look for stuff that is obviously out-of-place.

At the same time that you're looking for things to clean up and mend, you should be learning exactly how your team handles change.

Have they given you free reign to simply commit to the main branch? (Probably not!) Or do they want you to submit a pull request to GitHub and then request a review? And do they let you merge the change, or does someone else handle that?

Something else you should be taking notes on is exactly *how* your team curates the change log. In other words, how does your team format commit messages? And how do they *do* merges? Create a branch, rebase, then merge back? Cherry picking? What about multiple supported versions? Are you supposed to work against the latest and backport...or work against the oldest and merge forward?

On some teams, the format of the commit messages and the way branches are merged is vital to long-term maintenance. Find out if that's *your* team and observe exactly how they do things.

If this is starting to sound overwhelming, that's cool.

Learning *how* your team does commits may be more of a Month 2 or Month 3 thing.

And this doesn't mean you are *doing* the process. I'm just asking you to *become familiar* with how they are doing it. If you have an approved change, you can ask to do a pairing session to walk you through the process.

Often, your teammates will be excited to work with someone eager to observe The Way.

IF YOU'VE BEEN at your new position for at least six months, then that's great! Hopefully you are not only enjoying things, but effecting change.

Hopefully, the following list captures at least some of the things you're involved in:

- Learned how your team handles changes, whether it's new features or bugs
- Become familiar with how they use CI and work with it, not against it
- Gotten ramped up on how tickets are managed

Six months in, and you should be making changes. You may not be building the biggest stuff (yet), or maybe you are? However, what's more important than implementing the biggest features is learning the rhythm of your team.

Hopefully, you are learning how to bring value calmly and consistently.

They hired you for a reason. It wasn't to be combative, but instead to contribute. While implementing Big Stuff is cool, closing out a dozen tiny tickets with small stuff is always a good idea and will earn you some well-deserved allies.

We all make mistakes (myself included). If you have shown your ability to knock out bugs and deliver patches, teammates are willing to work through your mistakes. If you aren't pulling your weight, mistakes can turn against you much faster.

Again, focus on your team's processes and strive to contribute to it every day.

By the way, remember that daily journal we talked about earlier? Have you kept up with it? If not, then today is the perfect day to re-engage with doing that. It will only help you in the long run!

THE ANNIVERSARY of your new position rolls around. It might surprise you when it does! This is a time to celebrate.

And hopefully, it's also a time to push forward with providing even more value. And ideally, at this stage, you have built up a sense of the tools and apps you are assigned to.

But don't just settle for that.

Hopefully by now you'll have a stronger feel for the types of requests and bugs the community has been reporting. With that awareness, you need to start looking to the future, and asking, "What should we add that would really help the community?"

You can't always ask people what they want. They don't always know. This is Apple's mantra—building visionary products. Sketch out some ideas and discuss them with your team. This is the future, not just fixing bugs but instead figuring out where the app will go!

Also, you should be looking at ways your app interacts with other apps that you don't work on directly. As an example, when I was on the Spring team, it had multiple sub-teams, each working on various parts of the portfolio. And all these sub-teams worked with each other to ensure that the various facets of the portfolio worked together in harmony. You should seek to do the same.

And with a year under your belt, don't stop. Because you're just getting started.

Also, you may want to ask yourself where you are in your present journey...and write it down in your daily log. Yes, you can log daily tasks. But you can also capture thoughts and ideas, things that haven't fully baked. This can also be a place to start building your vision of the future!

Spring (https://spring.io) is popular for several reasons. In case you're unaware, Spring has a long history of solving problems, and that's one of the reasons people like it. Of *course* people want to use it. Some people don't use Spring, but it's not because the team doesn't fix stuff!

So what do you do if you're staring at a chunk of code and wondering, "What do I NOW??"

You're beleaguered from moving forward because something stumps you.

If you haven't felt that yet, don't worry. It will happen someday. ALL software developers run into this. And not just once. I certainly have. Many times.

That's why we need lifelines. One of the most powerful tools we as software geeks can have in our possession is a collection of fellow coders we can reach out to and ask for assistance.

Not only can seeking out help move us along...it's one of our best instructional tools available.

When something goes wrong, and we make it right, we learn in the moment. The *next* time we encounter something similar, we may solve it faster. Or perhaps we may spot it before it becomes a problem.

But asking for help is hard. It requires surrendering your

personal ego. A certain amount of vulnerability is communicated when you say, "I need help."

At a previous position, I once inherited a Java Swing app. It "worked" but was fraught with bad design practices, had horrible performance, and whenever someone hit the SAVE button, it would take twenty minutes to complete. I spent at least three weeks *simply trying to grok* the database operation behind that SAVE button.

There were no tests. No notes. Just a database view and the application's user interface.

I eventually figured it out. In building the view that served the application, the original developer had reached for other database views that had what he needed and joined them together.

In case you didn't know, joining views together in a relational database query is *bad*. The real query underneath ended up joining the same table *twenty times*.

After unraveling all this, I wrote a fresh query that did the same...and it went sub-second. I rolled out that patch to the users, and alerted them to this change, a change that probably saved them from thousands of hours of wasted time in the future.

The prior developer had worked *very* hard to implement it. But didn't pause and ask for help. Ever.

I don't know the story behind this person. Maybe they were too proud? Or maybe help was simply hard to find at the time? Who knows?

My point is that this instance has reminded me over and over, throughout my career, that it's usually better if I check my pride at the door and instead reach out for feedback and inputs more often than not. When things get a bit challenging, I reach out for help.

And if at all possible, don't work silently, in a corner, in isolation, avoiding all the opportunities to learn and grow as a software developer.

Anytime *you* are challenged with a situation, reach out for assistance. Anytime you're given a task, you can simply share what you are doing, whether it's with your instructor, your manager, or your technical leader.

Sharing what you're doing in a consistent fashion opens up the lines of communication and makes it easier to say, "You know that thing I've worked on for the past three days? I seem to have hit a wall. Can we Zoom for fifteen minutes where I share what's happening?"

The person you share with may not even say a word in the whole video chat. Sometimes simply enunciating the struggle you are facing can put the answer into focus.

My nephew who recently graduated high school (valedictorian, I may add), has called me late at night several times in the midst of coding projects. Most of the time, he simply talked while we were both looking at the same code. And he figured it out! I'd like to claim more credit, but I honestly did little more than say, "What's that?"

But even if you *do* need actionable advice from a colleague, it's okay. This is how software development *really* works.

None of us survive by working weeks on end in a secret room all by our lonesome selves.

Nope.

We thrive when we work together. Everyone has different talents. Different skills. Different things we have a certain nose for.

So what about persevering? When is it time to stop talking and start coding?

The truth is, there are times when no one else can help you. There are times when you need to try one thing, test it, alter it, test it, alter it again, and test it again.

Rinse. Repeat. And then perhaps three hours later, hang things up and go home for the day. Next day, resume.

It's true.

Pro coding is a career, not a hobby. Hobbies can be shelved at a whim. We are being paid to develop things, and sometimes that involves hammering until we make a breakthrough.

The struggle is knowing when to keep hammering vs. when to stop and get input. Sometimes, you can see incremental improvements even if you can't see the end (yet). This may be a good time to keep your nose down and keep the progress moving.

It's when you throw your hands and can't see any changes happening that you may need to reach for that lifeline of help.

And if you're looking for a rule of thumb, try to check in with someone else *at least once a week*. Whether you made huge progress or none at all, carve out time to share your situation with your manager or task leader at least once a week.

And over time, you will slowly but surely learn more about what your natural talents are.

One of my teammates is incredibly good with JVM-based stuff. We also have experts on some of the deepest, most intricate details of the Java programming language. My superpower is being able to hack my way through server-side dev-ops type stuff and bending CI systems to do my bidding.

Understanding what you can and can't do is a point of self awareness. For myself, I learned that while I'm great at relational databases and could write left outer joins and correlated subqueries in my sleep, my knowledge of the depths of JPA was much shallower.

When I worked on the Spring Data team, I was responsible for managing the CI pipelines for all fifteen of its modules. We went through many evolutions of that.

When I embarked upon another adjustment to our CI solution, I had succeeded enough times that I trusted myself to make it through. That experience had imbued me with confidence to get stuff done. But building up that muscle memory took time and effort. It didn't happen overnight.

Don't be afraid of the process of learning and building up your own set of skills. Having completed a programming course or some certification oftentimes grants you a beginner's level knowledge at best. Growing that into advanced or even expert-level skills takes time working on systems.

When we harness our own strengths and work with others, that's when we thrive. And that's when everyone thrives.

But one thing that can *help* you thrive is having the right structure in your code. Check out ***Rule 5: Structure Matters*** to find why.

RULE 5: STRUCTURE MATTERS_

There is something that underpins everything a pro coder does.

It's the code!

I once gave an annual review to a developer I was responsible for. This person had participated in many community-building events. Some at the company and some after hours. But his code-building talents were lacking.

I explained that community events don't really count. Only if two coders were equally tied and we're looking for a discerning difference does something like this come into play. To which I must add, is never. No one is ever exactly equal to another in coding ability for a given context.

Simply put, writing code beats all. Even developer advocates, whom many seemed to have impugned as not being "real" developers, have to be able to write code. In fact, advocates tend to be on their own to build the code needed to demonstrate products. Care to write code with no one backing you?

If you want a pay raise or a promotion, then writing knockout solutions that earn the company money is the golden ticket.

And so in ***Rule 5: Structure Matters***, we'll delve into techniques and tactics that can help you produce better code.

IF THERE IS one thing the Spring team is known for, it's their **code polishing**. A practice first begun by Spring founder Juergen Hoeller, this practice of spending extra effort to fine tune code formatting to ensure a *highly* consistent style has come to earn a rather unique name in the industry: **juergenization**.

The Spring Framework gets patch contributions all the time. It is with some frequency that Juergen would review the contribution and then respond with something like, "I appreciate your contribution. I took your patch and extended it to handle a more general situation."

The final version of the code that would actually join the Spring Framework would be written in a completely different manner, including renamed variables, re-ordered methods, and other aspects such that this contribution looked as if it had been part of Spring's code all along.

You may feel as if the code is what is **juergenized**. But actually, any contributor that makes lots of contributions begins to anticipate the changes the Spring team will apply. The naming conventions, the formatting, the order, the visibility settings. In time, contributors change their contributions to be more...Juergen-like.

And so, in truth, **juergenization** *really* refers to a change in the developer, not just their code.

That was quite a lengthy tale. So what's the big deal? Why is it so important?

Because maintenance over the long term is critical.

The Spring Framework first debuted way back in 2003.

That's a whopping twenty-one years at the time of writing. And because of the nature of entropy, that is, the tendency of things to fall into decay, if you don't exert deliberate energy to maintain order, the code will tend toward chaos.

This may sound like a dramatic exaggeration, but it's not. Software can actually swing faster into decay than many things. How?

As pro coders, we have the ability to make huge changes.

When I worked on the Spring team, I saw patches that would make monumental changes. And for something that should hopefully last at least another twenty-one years, the team simply couldn't swing the pendulum that fast.

It's important for any piece of software that is used by literally millions of developers supporting billions of dollars of industry to exert some measure of order. That's why the Spring team has always had a very strong policy regarding breaking changes. Essentially, no breaking changes nor any major upgrades of 3rd party dependencies were permitted in a patch release.

This is also why the Spring team would build using newer versions of Java but ensure the generated artifacts are in the format of the older version of Java that is the currently supported version. For example, back when the Spring Framework was Java 6-compatible, it had support built in for Java 7 features that were starting to emerge, like the newfangled **Optional** type.

This was possible due to certain chunks of code being wrapped in special runtime checks such that they'd *only* come online if Spring detected the presence of Java 7 (or higher).

Spring Framework APIs were protected at all costs such that upgrading to the latest patched version would *never* break a user. And ideally, upgrading to the latest major version won't break you either. This level of structural design isn't easy and requires firm commitment.

So when you sit down to do your coding, and you're contemplating adding a new class, or perhaps simply a new method, are you applying this level of analysis? Are you asking whether or not the change you're about to apply would break an existing user?

This type of care and consideration was something I wasn't even vaguely aware of early on in my career. Someone else owned that piece of the system that I worked on. But in today's world, fledgling coders are given bigger tasks a lot sooner.

And so it may land on your shoulders to evaluate if you're introducing something that is breaking!

In addition to managing breaking changes, there is also a strong likelihood that anything you add today, you'll probably be the one fixing it tomorrow should someone report a bug. Which takes us back to the benefits of investing time in code polishing.

As a 1st order effect, polishing code improves readability. It's well established that pro coders read much more code than they write or edit. Reading a chunk of code and being able to comprehend what it does is vital. So stopping to sift through a hastily thrown together solution and ensuring all the variables have suitable names and that it isn't really a hack but a solid solution, helps immensely.

As a 2nd order effect, a continuous effort of cleaning up, renaming, and striving for improved readability ensures that the *team* is working in a more harmonious nature. It's very hard to uphold such a team-wide status if one person is polishing while someone else on the team is just throwing in quick fixes and committing them.

This is why the Spring team adopted code formatting tools to help simplify the application of juergenization.

By the way, Juergen polishes his contributions manually, but last time I checked, the Spring Data team still uses the *Eclipse Code Formatter Plugin* on IntelliJ to automate spacing, import statements, line wrapping, and other aspects. However, things like naming, method ordering, additional spacing, and some other styling has to be done manually.

Here are some additional tactics you can consider when it comes to ordering parts of a class:

- Static fields should go first, followed by instance fields. That's because they are more "global" in a sense, being class-wide.
- The more important or relevant a field is to the class, the higher up it should go in the declared list.
- Each field should have its own line. Yes, you can declare multiple fields on a single line, but in general this works against you when it comes to readability and overall maintenance.
- Public, in general, goes ahead of package-private, then protected, and finally private. While you should rarely have a public field, a public constructor should go ahead of a package-private one. And the private constructors should be last.
- These tactics apply for fields, methods, anything.

There is a little variance in where to put those private methods. Since private methods are really there to serve non-private ones, you either shove the private methods all the way to the bottom of the class, or you instead put them right below the methods they serve. I tend to favor the latter!

Is this amount of thought about simply *ordering* the various parts of a class something you've never done? Ever seen a fellow colleague do this?

Trust me, it may feel costly at first to invest this amount of effort over ordering things. But it will pay off in the long run. When *every* class has this type of structure applied, it begins to support you.

A few years ago, I worked on a combined effort with engineers from Netflix, Google, Microsoft, and a couple other companies. The code was owned by another team, and I was providing the bits that supported our end of things.

And I immediately felt the lack of focus on simple code polishing.

The parts that I wrote were fine tuned. Some parts that were beyond my specific bits, but to which I helped with, I did the same. I chuckled when chatting with a fellow Spring teammate at that year's SpringOne conference. My teammate reacted, "So you *juergenized* it?" I nodded and smiled.

You see, when I was a part of the Spring team, I was part of a team who all relished that type of order.

If this is simply too abstract, too hard to grok, then I hope you are at least made aware of it.

Now there are some more detailed aspects to consider when it comes to properly structuring your code.

THERE's an age-old debate amidst coders: **spaces** or **tabs**.

Essentially, should all the whitespace between the text of your code be filled with spaces...or with tabs? Spring Boot co-creator Phil Webb once gave a talk where he posted the following picture:

A calm and measured response to this debate, which I must warn you can descend into anarchy more quickly than a Reddit thread, is to simply use what you want. Just pick one. Don't mix them together or else things will look horrible in the commit logs.

If you're a little wet behind the ears, you may be wondering, "What's the big deal?" How can something as simple as invisible whitespace lead to violent X threads?

As with most things that have strong opinions, there is actually a rich history to all this. At the risk of gross simplification, it goes a little like this:

1. People once used tabs, because the TAB key on the keyboard is meant to "bump stuff over" a bit.
2. But when viewed on certain websites (**cough** GitHub), every tab would be rendered into 8-spaces, a *ridiculous* amount of space for a single tab.
3. Hence, the developer community revolted and adopted spaces over tabs.
4. To avoid having to physically hit the SPACE key four times every time, IDEs adopted the

> convention of automatically transforming hitting the TAB key into inserting four spaces.

Despite this huge swing toward space-based whitespace, many teams have decided to remain with tabs (not just the Spring team).

There are strong arguments that tabs represent semantic distance. However spaces ensure that everyone is seeing the same thing.

But a new awareness surfaced that I wish to share.

Have you ever heard of an audio assistance tool? It's an application that some visually impaired people use. It literally reads content out loud. Could you imagine listening to "space space space space space space space space space space space space" compared to "tab tab tab"?

Some visually challenged coders have to shrink the text way down to single-spaced gaps in order to see it. Others have to make things a lot bigger. When you adopt tabs as the way to space out your code, how a given developer *sees* there is a preference setting!

It became a reality to me when I read a blog post by a visually challenged developer. Whenever he works on a new feature, he first reformats the code base such that he can see it. He makes his coding changes. Tests things out. Once completed, he converts the code base *back* to the spaces-based format.

Every. Single. Time.

I was already in the camp of "tabs = scoot stuff over," but had a rather indifferent opinion of others not wishing to adopt this.

Reading the struggle of this coder and feeling the angst I would go through were I to have to do this every day really swayed me. We really *should* adopt tab-based formatting to reduce the barriers to others making code contributions.

Now maybe this isn't your choice on the team you have joined. But at least I can offer you the same awareness I discovered to share with your team should the opportunity present itself (and perhaps you'll find this as compelling as I.)

ANOTHER ARENA where you can grow your skill set is using visibility settings properly. If you're not using Java or any other JVM-based language, then you can skip over this section.

Visibility settings may sound like something that isn't a big deal. But understanding the role of the visibility of classes, constructors, methods, and fields can step up your coding game.

The first thing to be aware of is that not *everything* your IDE chooses is necessarily the best. Usually, any IDE will default to making a new class **public**. But that actually shouldn't be *your* default.

Perhaps you're thinking, "So I should make classes **private**?"

No. And the answer isn't **protected** either.

The answer is the super secret ***default*** visibility. This means you have *no* visibility modifier. This is also known as **package-private**. Essentially, only classes in the same package can see this class. And that's the best place to start.

But why?

Simply put, any **public** code you release to the world can be consumed by anyone. If people start using your **public** class and build large systems on top of it, it becomes very hard to change. Removing it is even harder! Remember earlier in this chapter when we talked about not making breaking changes?

Making it **package-private** means the outside world won't have access to it...unless you choose to do so. While noodling out a new piece of code, making all of it **package-private** means you can actually release it and let it run in production. Then you are free to alter at will. No one outside your sphere of influence will be using it. At least not without some really hacky tricks!

Package-private will let you iterate across several releases. When I introduced the parser solution for HQL to Spring Data JPA, I kept everything **package-private**. This let us release something that was polished to our standards, but left us some room to make adjustments should we discover unforeseen issues after release.

When you make your code **public**, making sweeping changes becomes a breaking change that shouldn't happen until you make another major release. Or at least you *shouldn't* make such changes if you care about not disrupting your community.

The same can be said for constructors. Constructor calls are the entry points for other classes to first access your code. And so the initial visibility should be the same as your class: **package-private**.

Constructors are a little different from classes. That's because one class can have multiple constructors. For different starting scenarios, you can write a different constructor. And if this one is being accessed from outside the class, make it **package-private**. But it's possible to extract out some common construction behaviors that your various constructors need. Hence, you can have additional ones that are **private**. A **private** constructor is one used by another constructor.

Now that's a *lot* of talk about class and constructor visibility. If this is perhaps 5-10% of a class's code, why invest so much time and energy?

Because *real* object-oriented design is centered on this.

Some coders have a jaded perspective of isolating things. *Polymorphism* and *subclassing* aren't quite the magic we were all promised twenty years ago.

But a much firmer attitude of *only* letting outside classes see *exactly* what they should see is the sign of a serious, seasoned pro coder. Now it's quite possible that classes outside this class's package need access, thus forcing you to make the class and at least one its constructors **public**. But don't rush to that position.

It's always possible to wrap the class with an interface type and have it constructed by some context-aware factory. Perhaps a concrete example is in order?

Again pointing to Spring Data JPA's HQL parser, I had introduced an **HqlQueryParser** This class parses HQL queries. And it's **package-private**.

```
/**
 * Implements the {@code HQL} parsing operations of a {@link JpaQueryParserSupport} using the ANTLR-generated
 * {@link HqlParser} and {@link HqlQueryTransformer}.
 *
 * @author Greg Turnquist
 * @author Mark Paluch
 * @since 3.1
 */
231 usages
class HqlQueryParser extends JpaQueryParserSupport {

    1 usage
    HqlQueryParser(String query) {
        super(query);
    }
```

The only class that needs it is **JpaQueryEnhancer**, found in the same package. It has a static method that's invoked from a factory *only* when we detect Hibernate on the class path.

We were able to roll out this new feature to our user community with zero disruption because it's behind our already-existing APIs.

Which leads us to the point that how you create and invoke your code is at least as important as what the code itself does!

A rather common pattern is to make all of a class's fields **private** and only grant access through **public** getters and setters. In fact, most modern-day IDEs include the ability to auto generate these accessors. Oftentimes that's fine.

But be aware that you can create *virtual* properties. You can always create additional getters that are built from other fields. For example, imagine your class had a **firstname** and **lastname** field along with **getFirstname**() and **getLastname**() methods.

A natural extension would be to add a **getFullname**() method. This method would be easy to implement. Simply combine the **firstname** and **lastname** fields with a space in the middle. And if you make it **public**, many tools and handlers will treat it like another property!

Jackson, one of the most popular JSON serialization libraries for Java, will render an extra field to the rendered JSON output as if that field were there the whole time. You can also implement **setFullname**(), and have it split up the inputs and set the real fields when deserializing a JSON document.

Properties can also be hidden from Jackson by switching them to **private**. If it's needed by other parts of the package, you should consider **package-private**. Jackson also has annotations to help steer JSON production and consumption, but we won't delve into that.

Class design is important. And the visibility of your elements is a vital aspect that shouldn't be overlooked. But that's not the only thing pro coders must pay attention to. The way they structure commits is also of keen importance.

If there's one thing you must do as a pro coder, it's to commit changes to your code base. That's because every corporate entity uses version control. Every non-profit coding institution uses version control. Every university uses version control for their operational systems.

And if they don't, run because it's the standard and anyone that's not using it yet is working in the Stone Age.

So what, exactly, are the most vital things you need to be putting in your commit messages? How many commits should there even *be* for a new feature that you're working on? Should it be a historical record that captures every step of the path you took to implement something? How does this correlate with tickets used to manage changes to the code base?

Let's jump in by first saying that the most important thing you should put into any commit message is what you did. Maybe that sounds kind of obvious? Let's dig in first by looking at the example below where I first introduced the query parser mentioned in the previous sections:

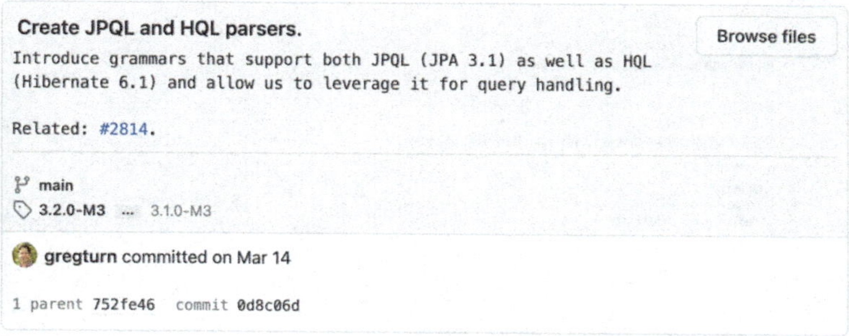

The title of this commit clearly describes what happened. It's short and to the point. The body of the commit message mentions vital details like what versions of JPQL and HQL this encompasses.

What's also valuable is what this commit *doesn't* have. It

doesn't discuss how the decision was made, nor does it go through all the activity taken to implement this enhancement. Instead, all those superfluous details are left out.

It also doesn't waste time explaining what **JPQL** and **HQL** stand for. We're all adults here. Let's just assume the person reading this commit message knows the technology. If they don't, they can look it up!

If you're wondering where we should track the discussion and design decisions, then continue reading this section. But first, let's tackle another type of commit, a bug.

The commit above captures a planned feature. For an end user who files a bug report with less than clear details and you have to do some investigation to unearth what the real issue is, does any of that research need to be in the commit message?

No.

Here's an example of just such a ticket:

BadJpqlGrammarException no viable alternative at input for reserved keyword: cannot escape #3056

Closed multanis63 opened this issue on Jul 7 · 2 comments

multanis63 commented on Jul 7 · edited

Tip ···

Hello,

Since 3.1.1, the following entity / query is not working anymore:

```
@Entity
@Table
public class TestEntity {

    @Id
    @Column(name = "ID")
    private Long id;

    @NotNull
    @Column(name = "TEST_TYPE", nullable = false)
    private String type;
    }
}
```

```
@Query(value = "select te from TestEntity te where te.type = :type")
List<TestEntity> findByType(@Param("type") String type);
```

Result: org.springframework.data.jpa.repository.query.BadJpqlGrammarException: Line 1:34 no viable alternative at input 'te.type'; Bad JPQL grammar

It's great that they captured the precise exception encountered, but the rest of the title is rather vague, probably because they aren't quite sure what's happening.

Before we get to the actual commit that fixed this bug, let's dig in a little more on the initial ticket. For starters, if there isn't much there to point you in the right direction, then it's perfectly fine to add a comment asking for more details to help figure out how they stumbled into this issue.

In this example, they already captured the scenario quite nicely. There is an example of an entity class along with a query method. All of that can nicely be transformed into a test case that reproduces the issue. Request for more detail not needed!

Based on all this, let's fast forward to the commit that was rolled out to fix this issue, as shown below:

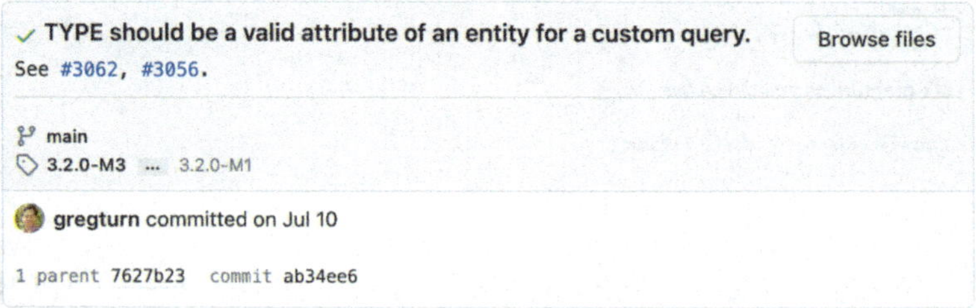

Let's contrast this commit message against the issue originally opened.

- The title of the commit message concisely states what was fixed.
- The body of the commit message mentions related tickets. It should be pointed out that when you prefix an issue number with a pound sign, GitHub will automatically turn that into a link to the ticket or pull request.

- Due to the small scope of this issue, there is no real need for anything more than the title and ticket links. But what if the commit had a lot more to it?

For every commit you create, you must decide whether or not to include more detail after the title. The point of the commit message is to present the gist of what was done, not a long, winding path including every tweak you made along the way.

Let's go back to the parser commit we looked at earlier.

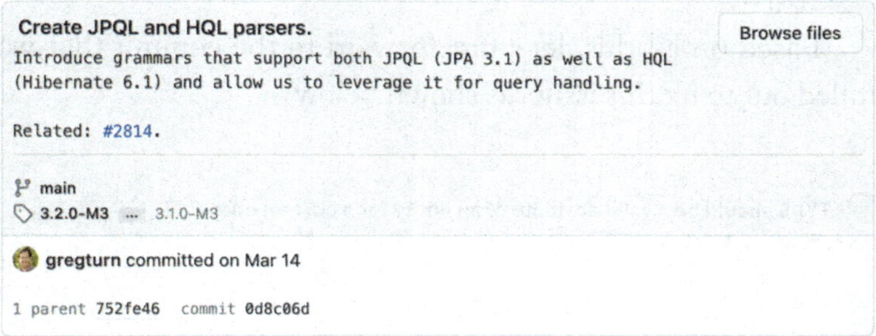

Notice how it mentions precisely what versions of JPQL (JPA 3.1) and HQL (Hibernate 6.1) it covers? It also states that this was implemented to handle queries.

All of the various conversations that you had while hammering out the details of the patch, whether between teammates, community members, and other involved parties, should certainly be a part of the ticket's conversation. But not the commit message itself.

The commit message should lay out what was wrong with the code and what you've done to resolve it. In other words, someone *should* be able to read your commit log and have one hundred percent clarity on how the code evolved.

But if they want to see *why* the code evolved that way, then they'll have to go read the relevant tickets. *That* is where all the discussions should live.

Now before we discuss what to *never* put in a commit message, it's important to link back to the original discussion. The last thing every commit message should include is one or more ticket references. In general, there should be at least one ticket, which precipitated this change. If you'll notice, all the commit screenshots shown so far include a link to relevant tickets.

If the commit is connected to multiple tickets or there's different related tickets that circle around the same issue, you can mention those as well.

Simply put, we're trying to make it where anybody that finds this specific commit can read the description, read the code changed, go into the ticket logs, and understand what happened.

As one last example, look at the following commit from Spring Modulith, a new Spring project that helps you build better monolithic applications:

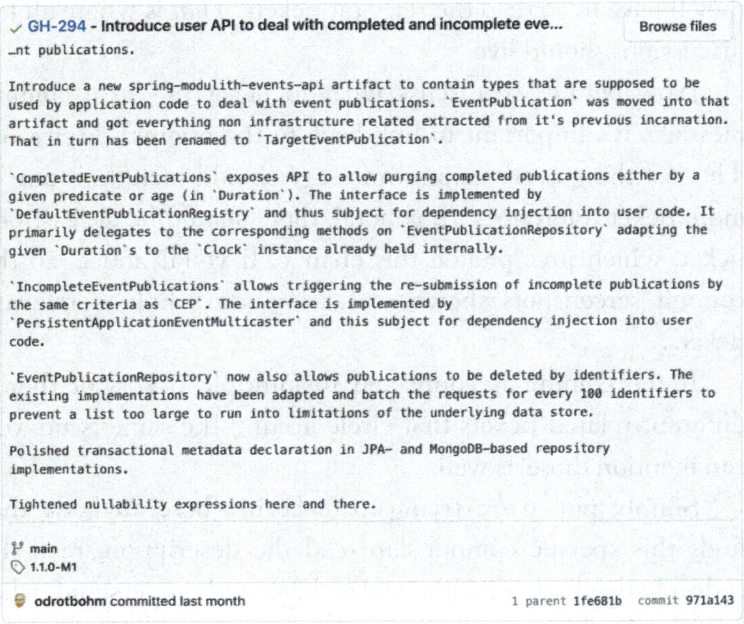

✓ GH-294 - Introduce user API to deal with completed and incomplete eve... Browse files

…nt publications.

Introduce a new spring-modulith-events-api artifact to contain types that are supposed to be used by application code to deal with event publications. `EventPublication` was moved into that artifact and got everything non infrastructure related extracted from it's previous incarnation. That in turn has been renamed to `TargetEventPublication`.

`CompletedEventPublications` exposes API to allow purging completed publications either by a given predicate or age (in `Duration`). The interface is implemented by `DefaultEventPublicationRegistry` and thus subject for dependency injection into user code. It primarily delegates to the corresponding methods on `EventPublicationRepository` adapting the given `Duration`s to the `Clock` instance already held internally.

`IncompleteEventPublications` allows triggering the re-submission of incomplete publications by the same criteria as `CEP`. The interface is implemented by `PersistentApplicationEventMulticaster` and this subject for dependency injection into user code.

`EventPublicationRepository` now also allows publications to be deleted by identifiers. The existing implementations have been adapted and batch the requests for every 100 identifiers to prevent a list too large to run into limitations of the underlying data store.

Polished transactional metadata declaration in JPA- and MongoDB-based repository implementations.

Tightened nullability expressions here and there.

main
1.1.0-M1

odrotbohm committed last month 1 parent 1fe681b commit 971a143

This commit message is slightly different from the previous ones. For starters instead of using the pound-prefix issue reference (#294) it instead uses GitHub's other issue reference mechanism (GH-294). This, too, will let you link to issues. And for this project, the critical issue is mentioned in the title of the commit message.

But like the prior commits, it has a clear title that tells us what this change does. It introduces an API to deal with two specific types of events.

If you continue to read the rest of the commit message, it is much more detailed than the earlier commits we have already seen. While longer, it is certainly easy enough to read, thanks to the committer breaking things up into bite-sized paragraphs.

It also adopts a writing style often seen in news articles: it has the most critical information at the top with extra details further down. Compared to a novel where the author wants to draw you to read to the end, if you were to stop reading after the first paragraph, you'd have learned the bulk of what this commit does. It's not critical that people read every word of a commit message.

Now that you know what to put in a git commit message, it's time to answer that age-old question, "How many commits should I create for a given feature or patch?"

The simple answer is **one**.

That's right. For a given ticket you should have just **one** commit.

There are some interesting exceptions. For example, to comply with coding practices that the team has adopted, your tech lead may go into the code that you just completed and polish things up with an additional commit. He or she may apply some changes like changing the import statements or removing superfluous spaces, and it's okay for that to be separate from your commit.

The reason for only 1-2 commits is in the event someone is reading the history of your project. There is no need to make them read three, five, or a dozen commits to get the whole picture for a single change.

Now don't confuse this with daily workflow. While working on a branch, it's perfectly fine to have multiple commits. Many pro coders will have a flow that goes a little like this:

- "verified the problem"
- "fixed main use case"
- "fixed huge corner case"
- "added more tests"
- "fixed half of them"
- "fixed the rest"
- "added one more corner test case"

- "fixed it"
- "polishing"

That's nine commits and that's fine. There is no reason at this stage to be focused on writing the perfect commit title. Instead, the focus is on exposing the problem and fixing it! That's why I like to call these **in progress commits**.

But somebody that reads your code after the fact doesn't need this awareness of your workflow. That's why when your solution is all coded up, you can simply squash all those **in progress commits** into a **single commit**.

To do that, use git's **interactive rebase**. Git will show you a list, oldest to newest, top to bottom, like what's shown below:

```
pick 26533a1cf Create branch.
pick b77cb70ad try
pick 4c5b782bb Make JpaSort.unsafe work.
pick ccd68406e Multiple literal args.
pick 0880f9def more
pick 58b659245 done

# Rebase 867de7d8c..58b659245 onto 867de7d8c (6 commands)
#
# Commands:
# p, pick <commit> = use commit
# r, reword <commit> = use commit, but edit the commit message
# e, edit <commit> = use commit, but stop for amending
# s, squash <commit> = use commit, but meld into previous commit
# f, fixup [-C | -c] <commit> = like "squash" but keep only the previous
#                    commit's log message, unless -C is used, in which case
#                    keep only this commit's message; -c is same as -C but
#                    opens the editor
# x, exec <command> = run command (the rest of the line) using shell
# b, break = stop here (continue rebase later with 'git rebase --continue')
# d, drop <commit> = remove commit
# l, label <label> = label current HEAD with a name
# t, reset <label> = reset HEAD to a label
# m, merge [-C <commit> | -c <commit>] <label> [# <oneline>]
#         create a merge commit using the original merge commit's
#         message (or the oneline, if no original merge commit was
#         specified); use -c <commit> to reword the commit message
# u, update-ref <ref> = track a placeholder for the <ref> to be updated
#                       to this position in the new commits. The <ref> is
#                       updated at the end of the rebase
#
# These lines can be re-ordered; they are executed from top to bottom.
#
# If you remove a line here THAT COMMIT WILL BE LOST.
#
# However, if you remove everything, the rebase will be aborted.
#
~
```

The first commit I'll talk about a little further down in the section. The second one is where I started working on the issue at hand. Additional commits involve more test cases and solutions.

When you're finally ready to deliver your results, simply go to each commit (after the first one) and replace **pick** with **squash** (or just **s** for short) and then exit.

Git will take each commit and "squash" it into the commit above. If all are squashes, then all will get folded into that first commit. When that is done, you will also be presented with a chance to rewrite the commit message.

Use the tips we just discussed and rewrite that commit message to capture what changed!

What if you don't like what you wrote for that final commit message? Simply rewrite it using **git commit —amend** (notice the double dash). I do this all the time until I feel like the entire message is crystal clear to anyone.

The point is to make your multi-day task of fixing the problem look like a single change to the codebase, as if you sat down and did it all at once.

I should mention that some people don't like this. "That isn't what happened" is a common criticism.

This may sound strange at first, but it doesn't matter what happened. At least not from the perspective of *how you spent your time*.

What *does* matter is what happened historically *to the code*. What happened to the code is you implemented a change that includes test cases. The order of how this happened isn't relevant. Don't make the people reading your record of changes read a dozen commits.

As I said a few paragraphs back, there are a couple exceptions to this 1-commit rule.

If your manager or team leader, just before merging to

main, decides to pick up your branch and apply a little extra polishing, that's fine. We described that earlier in this chapter as **juergenization**.

In some situations you may not be authorized to merge to **main**, or you may simply be told on this assignment you don't merge to **main**. Instead, bring your branch to your team leader and your team leader will then do any final polishing. They may either merge the work to **main**, or they may give you feedback and ask you to do a handful of things.

Don't worry!

These are normal, healthy actions used by any team. If somebody is polishing your efforts, don't take it personally. This is how collaborative software development works.

There's another exception that I hinted at, shown in that interactive rebase up above. The Spring Data team actually performs something called branch builds where a given branch is actually published to our artifact repository for downstream consumption. Sometimes they need to test a branch in another app or in a downstream project while it's still being worked on.

Imagine the current version was **3.2.0-SNAPSHOT** and one of them started a new branch. Let's pretend the ticket being worked was #3172 in GitHub. The Spring Data team will actually create a commit and bump the version to **3.2.0-gh-3172-SNAPSHOT**. That way when the CI server builds it, they would actually stage that version in their snapshot repository. This lets them avoid overriding the actual **3.2.0-SNAPSHOT** that others may potentially be using.

They certainly don't want those two versions competing with each other so they'll create a **preparation commit** that preps the branch for working even if it's for a one-line change. They even have a script to simplify the process!

In fact, you can see part of the commit at the top of that interactive rebase right here:

```
commit 26533a1cf42a5f6fd4ae6ea746de337ecc47bc7e
Author: Greg L. Turnquist <gturnquist@vmware.com>
Date:   Tue Sep 26 16:15:04 2023 -0500

    Create branch.

diff --git a/pom.xml b/pom.xml
index 94542470a..ee0112a07 100644
--- a/pom.xml
+++ b/pom.xml
@@ -5,7 +5,7 @@

        <groupId>org.springframework.data</groupId>
        <artifactId>spring-data-jpa-parent</artifactId>
-       <version>3.2.0-SNAPSHOT</version>
+       <version>3.2.0-gh-3172-SNAPSHOT</version>
        <packaging>pom</packaging>

        <name>Spring Data JPA Parent</name>
```

Their script creates a commit where the only thing happening is switching the version of this project to that branch build version **3.2.0-gh-3172-SNAPSHOT**.

Then they do all the work.

When it comes time to do that rebase I talked about earlier, they'll squash everything into the *second* commit where the work really started, like this:

```
pick 26533a1cf Create branch.
pick b77cb70ad try
squash 4c5b782bb Make JpaSort.unsafe work.
squash ccd68406e Multiple literal args.
squash 0880f9def more
squash 58b659245 done
```

The second commit is where the work really began, and I was trying to simply recreate the problem. From there, I pushed through making more use cases work. Notice how the third commit and beyond are flagged to be squashed? Upon completion of this rebase, there would only be the first one and a new second commit.

In case you're wondering, the Spring Data team then cherry picks that second commit into the main branch (using **git cherry-pick**), leaving behind the commit that created an alternate version. This is not a common tactic, not even amongst other Spring portfolio projects, but hopefully you now understand the reason they do this. And even more importantly, that ultimately, they still uphold the 1-commit per feature paradigm that I mentioned earlier.

If you're still wondering what you should *never* put in a git commit message we're almost there, but first we need to answer some questions. What about trunk-based development? Or git flow? Or some of these other paradigms that people are using? If you haven't heard of them, they're not that complicated.

There are other approaches, like trunk-based development that talks about working directly off the **main** branch as well as Git Flow that involves branches other than **main** including **develop**, **feature**, **release**, and **hotfix**.

No matter what paradigm you pick, be it trunk-based, Git Flow, standard Git, or something else, the concept of squashing multiple commits into one still stands. You can be a trunk-based or a Git Flow-based team and still squash all your efforts into one easy-to-read commit to make it easier for others to consume your changes.

Simply put, there is no need for someone reading the record of code changes to re-read your entire experience of bringing about the change. And if a colleague polishes up your changes in a separate commit, it's okay to keep that separate. One commit for functionality, one commit for styling.

And always always *always* link back to the relevant tickets!

Throughout this section, I've made reference to 1-day or 3-day efforts. But what if you do something that takes three months? What if someone submits a pull request with two hundred commits?

I ran into this exact scenario several years ago working on a contribution to Spring HATEOAS. Whether we're talking about a really long period of time or a lot of commits (or both!), there are some inherent risks that must be dealt with.

For starters, there is a huge risk if you do not rebase your work frequently against **main**. You should rebase at least once a week or even every other day. If you don't you're going to run into a *lot* of changes you have to absorb.

That's because under the hood, rebasing involves grabbing the latest changes on **main** and reapplying your changes as if they were patches to the latest version of **main**. And when I say reapplying your changes, I mean each and every one of those commits on that pull request's source branch.

If you have two hundred commits, each one must be reapplied. And the more drift there is between main and your branch, the harder it will be to carry that out.

Which is why you should probably *first* squash everything into one commit.

Why?

Because anytime you rebase against **main** to catch up, every commit on your branch is put through the process. With two hundred commits, we're talking about two hundred mini-rebases. This is the best reason to consider squashing a long stack of commits into one, and doing it somewhat frequently.

This is also known as **reintegration**, because you're reintegrating the updates of **main** into your branch.

Now you should keep moving at your pace, whether that means every other day or at least once a week. Carve out enough time for your tasks, as they may turn into hours of work. Rebase everything and squash it down. This practice is going to save you the next time you have to do it.

This conversation should also include dealing with spaces or tabs in the comments. Whatever you're doing, be mindful not to alter functionality and make formatting changes in the same commit. Major changes in spacing/tabbing is essentially a style change, and we already discussed the benefits of making that happen in a separate commit.

It's also important to emphasize that when you mix functional changes with formatting changes, and if the formatting changes are extensive, it's almost impossible to read the diff report on GitHub and discern the functional changes.

Are you dealing with that in a given submission? Then make that reformatting separate with its own commit. Apply your changes from tabs to spaces, or vice versa. Afterward, commit that change, merge it to main, and close out the day. Don't forget to write a ticket for that.

Afterward, move to the actual issue you're solving and rebase to that.

Which brings us to finally mentioning the things that *never* should be put into your change log—the war story behind it.

We've all had less than stellar coding experiences. Hopefully, they aren't all the time. Nevertheless, there are times where you are tempted to pour your heart out in the commit log.

Well think twice and shrewdly consider *not* putting your war story in the commit message.

Whether the grief you're feeling is coming from a fellow teammate, a community contributor, or the boss, don't put it into your commit message. The record of your code should reflect how it evolved. Period. End of sentence.

It doesn't belong in the history. And it certainly doesn't belong on social media.

No matter how justified you feel, going to Reddit or X to complain about your manager is not a solution. Communication between professional coders demands a certain standard, a certain decorum that can be summed up in the Pro Coder's motto: **Be Professional**. We'll discuss that in more detail later in this book.

Now there are other key things to improve your output that we'll cover next in ***Rule 6: Test What You Build***.

RULE 6: TEST WHAT YOU BUILD_

Not only are we summoned to write code. The code we write must work.

Way back at the dawn of my career, we had entire test teams tasked with running reams of printed test procedures against feature branches before they were merged into the official code base.

I'm here to tell you that the state of testing has morphed significantly over the subsequent two decades.

Test-driven development has risen. Servers dedicated to running test suites (**Continuous Integration** or **CI**) are standard practice. A legion of assertion toolkits make it possible to elegantly and succinctly capture test scenarios. There are even tools like **Testcontainers** to launch databases, message brokers, and other middleware products without having to install such products.

And the most popular toolkits of today, like Spring, come loaded with test-oriented utilities to spin up application instances in various configuration modes in order to drive test scenarios.

Long story short, if you're new to this industry, you must be made aware that testing is commonplace. With ***Rule 6: Test What You Build***, we'll see testing's various dimensions.

THE FIRST THING to grasp is how vital it is to test *everything*. In fact, there are test coverage tools in most IDEs that let you not only run your test suite, but identify which lines of code were covered and which ones were not.

One thing to be cautious about is how this can lead you down a faulty path.

It's quite easy using these code coverage tools to spot untested lines of code, but that isn't the whole picture.

Software isn't simply a series of lines. In fact, it's possible to cover the same line multiple times for multiple reasons. Perhaps an example is in order?

When I added the custom ANTLR-based parser to Spring Data JPA, I was working off the JPA specification. This document not only had lots of parser rules already written...it also had LOTS of fragments of standardized queries I could use as examples!

And I made a point to capture every single query fragment from that document as a separate test case.

These examples certainly touched some of my query parser's lines multiple times. And that was okay. A working proof of every example fragment of the official query language working in our system was vital to endear confidence. I even found errors in their examples using this!

What's perhaps not clear is that you *don't* have to write subsequent test cases *only* for the lines you didn't touch in the previous test case. If you craft multiple test scenarios, and they hit the same lines, the hope is that adding them all results in comprehensively covering a much wider area of code.

In fact, if you focus future test scenarios only on lines not previously covered, you will find yourself backed into a corner. Instead, use the lines of code not covered to give you an area to aim at. Write your test case. And if it happens to partially overlap code covered by another test case, that's fine. As long as over time you are expanding the coverage, then you're making solid progress.

Another example of thorough testing that isn't mentioned very often are pass/fail test scenarios. What am I talking about? It's easy to ask "what is this code supposed to do" and then turn that into an automated test case.

But what about the failing paths of logic? A RESTful API controller that ingests integers may not properly handle floats, characters, or other data types. You should indeed write test cases that feed not just good data to the system, but bad data as well.

The concept of pass/fail test scenario especially extends to security-based testing.

As an example, Spring Security has some powerful test tools that let you easily verify users with proper credentials as well as users with invalid or non-existent credentials.

Every successful path from a security perspective should include a counterpart test case for a mock user with insufficient authorization as well as a different mock user with no authentication.

NOTE: In case you didn't know, **authentication** is proving who you are. **Authorization** is what you're allowed to do. For example, people may be able to visit a bank's website and read about general information, but they would have to *authenticate* to access information only shown to customers. And they should only be *authorized* to access their OWN account and no one else's.

The type of analysis where you seek comprehensive coverage of scenarios, specifications, and security extends far beyond simple line-based coverage analysis.

In fact, focusing too much on coverage reports can drive you to write brittle test cases. I mentioned this in the *Python Testing Cookbook* back in 2009, but essentially, if you aim at writing another test case that covers *just* the untested lines, you may end up writing sloppy test cases that are hard to maintain.

Instead, think of complete, practical scenarios that will intersect with uncovered parts.

If you've heard of various testing strategies such as unit testing, black-box testing, white-box testing, integration testing, smoke testing, or some other types of testing, maybe you're confused that none of those have been mentioned. I'll cover them in the next section so you can understand exactly what role these various test strategies play.

THERE ARE countless books written about testing. Some of them document various flavors of testing including **unit testing**, **integration testing**, **slice-based testing**, **smoke testing**, and more. There are multiple reasons to choose a particular testing approach, and software developers debate the veracity of various approaches. But one thing is more critical than anything.

Write the test!

I once picked up a colleague's application because they were rolling off the contract. The application had a small set of test cases, nothing extensive. But apparently, this app wasn't operational. Over two years had been spent developing this tool for processing monthly invoices from telecom providers. Yet despite all this effort, not a single invoice had been processed with this tool!

I vowed then and there to *not* write a single line of code without first having a test case to cover it.

For this app to process invoices, it performed many spreadsheet-type functions. As a natural consequence, the first thing I invested time and effort in was figuring out how to make an automated test case ingest an invoice.

It was cryptic. It was challenging. It took me a couple weeks to figure it out.

But I did.

The invoice had to be in a specific location. And the results ended up in the database. Nothing between those two parts of the app let me "check" the results short of writing a database query.

To make things worse, the test case took a couple minutes to run.

But I got my arms around it. And from there I moved forward.

It took weeks to build up additional test cases, to verify the smallest changes. And to be honest, I didn't have time to think about whether this was a **unit** test or an **integration** test. The important thing was that this was an *automated* test.

Over the next several months, my test armada grew to requiring thirty minutes to run.

While time consuming, the presence of this test suite was vital. It built up my confidence as I slowly learned this app's internal workings. If I hadn't had that long-winded automated test harness, I would probably have never deduced that this app contained an unneeded abstraction.

Armed with such awareness...I ripped it out. And the tests sped up. A little.

The fact that I had built up a huge safety net and been able to remove something that was unneeded gave me the courage to continue pushing the envelope. I started pulling out other unneeded and unused features.

After about six months of having taken on this obstinate app, my manager asked for a status report. He wanted to know if it was going to work. Would the finance team be able to start using it to process monthly invoices instead of doing things by hand?

I offered to give a demo, but needed about three weeks to prepare.

I switched from writing test cases that covered specific features and bugs to instead capturing end-user scenarios.

These are sometimes called **acceptance** tests and are incredibly valuable to illustrate an application's features as it's meant to be used.

My test cases and test methods had *very* long, detailed, and descriptive names. It makes sense to capture the intent of a given test case in its name. These verbose test names can tell you what's wrong when one fails, hopefully putting you on the path toward fixing things much faster.

My scenario-based test suite read like a user clicking through a series of screens, but instead was invoking the application's various APIs.

I walked into that demo meeting with my test procedures printed out. I ran the application, using the test scenarios like prompts for clicking on the screens.

And it worked. Perfectly.

I didn't veer away from my test scenarios.

And it was in this meeting, after my demo, that I learned that six people before me had worked on this application. And that they'd actually spent millions in time and effort and had nothing to show for it over the previous two years, as mentioned earlier in this section.

Until that day.

I was stunned to say the least. The department that was supposed to use this application was eager to start, so we moved into operational mode.

We leapt forward at warp speed. With a testing philosophy of "no feature or patch without a test" firmly established, I began working with the users directly.

Every time they had a request, I captured a new test case. This is the whole point of **acceptance** tests.

Occasionally, the test suite's run time would creep up to over an hour, sometimes longer. And I'd switch from enhancements and bug fixes to cleaning out junk and removing cruft. Whatever it took to speed things up.

Once I squeezed test execution back down to thirty minutes, I'd say "good enough!"

A lot can be accomplished when we establish a threshold of "good enough" and don't push for purity.

But something else was happening at the same time.

At first, the point of the tests was to learn for myself how this application operated. Somewhere along the way, I had learned the ins and outs of this application. And so the test cases instead became the means through which I could decouple its backend processing from its user interface.

Eventually, I was having weekly meetings with the users in a conference room. We'd brainstorm new scenarios to use this tool. I'd capture each scenario in a PowerPoint slide. After the meeting, I transformed each slide into a separate test case and implemented it.

And it was fun!

I had never carried out such a fun cycle of meeting with

users, capturing needs, implementing them, rolling them out, and seeing their reaction. Simplifying things for others is truly a wondrous feeling!

This was how software design and development was *meant* to be.

And for such a tiny tool used by maybe six people at the most, it was one of the most satisfying things I ever did.

But it wasn't easy.

Spending weeks in the early days to get a single test case to even work in an automated fashion took resolve and commitment. To set aside simplistic blog-based opinions regarding "should I write unit or end-to-end integration tests?" and instead focus on making something work required a certain amount of personal confidence.

It also required commitment to my plan of writing those test cases.

I once tried to implement a new feature that was pretty big. I didn't run the test suite for at least three days. And when I did, half the test cases broke.

I tried to get them back to working status for a couple days. I finally gave up, threw away all the changes, and went home for the weekend.

I was saddened. But I also knew that was the right decision.

I had gone too far and my safety net screamed at me.

The following week I began anew. Only this time, I ran the test suite more often. And I resolved the issue. That was one of the hardest lessons I'd ever learned. Sometimes it's okay to throw away your efforts and start over.

My decision to write a test case for every patch paid off.

More importantly, it empowered this team of people from the finance department to do *their* job with ease and confidence. That was the *real* joy I felt working on that application.

I was there to help them. And when you help others and

are eager to build new features that simplify their lives, then the arguments over **unit** vs. **integration** or **test-first** vs. **just write the test when it makes sense** kind of melts away.

When you set your ego aside and instead help your users, that's an opportunity for genuine deep-seated satisfaction.

We're here to help others. That's what this is all about.

Every test case we write represents the user community we are serving. It makes sense to capture a proper test case that will always ensure the system behind it acts as our community needs it to.

And the entire suite of tests we build are like a group of our users picking up the system and giving it a rough shake.

With that amount of confidence, we can happily hand them a release.

And something that's key in any pro coder's repertoire is the ability to take releases to production. We'll explore ***Rule 7: Build Things with Production in Mind*** to cover what this means.

RULE 7: BUILD THINGS WITH PRODUCTION IN MIND_

Josh Long, the Spring team's first developer advocate, has often stated that his favorite place in the world (software-wise) is production. After all, what's an app that isn't in production?

Why?

This is where applications are meant to be. This is where they're performing their true purpose. And I may be so bold to add on my own reason why this is true, production is the real crucible for application behavior. Things break, databases get overloaded, networks misroute packets, and users click the wrong buttons. It's in production that we learn what an application is missing such that we can enhance it to truly serve others.

An application that crumbles into dust in production, simply put, isn't ready for production.

That's why we must embrace ***Rule 7: Build Things with Production in Mind***. It's nothing short of critical. In this chapter, we're going to delve into strategies that can turn our focus in the right direction.

WE ALREADY LEARNED in ***Rule 6: Test What You Build*** that it's vital to write tests for every feature and bug you come across. But tests alone aren't enough to actually carry us to production.

The next thing to do is embrace **continuous integration** or **CI**.

Continuous integration begins with taking every change you push to GitHub and running your entire test suite against it. The CI server can then report either a status of GREEN if it passes, or a status of RED if it fails. Doing this requires the adoption of automated testing, which we discussed thoroughly in the previous chapter.

If you're wondering why it's necessary to test every single change made to the codebase, even if it's just a single line, let me share with you a little bit of history.

In the olden days, we used to wait until right before making a release to do such verification of builds. Why? Because testing back then involved human involvement. We didn't have automated testing, but instead our test engineers would walk through stacks of printed test procedures. It was time consuming and if I'm being honest, prone to error.

But that wasn't the only issue.

Due to it taking so long to run these tests, releases weren't made weekly or even monthly. Instead, quarterly was the fastest I'd see a release get scheduled.

So what happens when you haven't released anything in three months and you haven't even run your full spectrum of tests?

Something would be broken. Sometimes multiple things.

The developers would have verified things within the scope of what they're working on. They'd have verified local behavior.

The most likely place where errors would creep in would be at the edges, the places where their stuff might be integrated with other components.

It was insidious. A developer may have fixed one feature and verified that it worked with the related component. But they weren't verifying that *all other* integration points weren't impacted.

Hence, the time when *everything* would get tested while preparing for a release would be the time of **system integration**.

And this is when we'd unearth stacks of issues that had to be resolved before making a release.

This was such a well-documented situation that integration testing and release preparation were formal parts of our schedule!

And this is why automated testing opened the door to an increased level of code quality.

Imagine that instead of waiting three to six months to discover some lingering bugs, you instead found out tomorrow?

A CI server that runs everything, whether it's your changes or not. The moment it breaks, the odds that the culprit was committed to the baseline in the last twenty four hours is quite high.

The sooner you can detect breakage, the sooner you can divine what went wrong and fix it. Or at the least, pull your change out of **main** and work on it some more.

And that is the first rung of **continuous integration**.

You see, as valuable as it is to *break* the build, it's even more valuable to *catch* such breaking changes before they become a part of the build.

Another aspect of **continuous integration** is a release-oriented team. That is, a team that not only makes changes on their own developmental branch, but are frequently updating their copy of the **main** branch and updating their own development branch to absorb these changes.

Instead of the way we did it back in the day where you cut a branch and then get to work, never seeing all the changes subsequently getting merged to **main** while you work in your own world, **continuous integration** demands that you are constantly integrating **main** back into your branch.

On Git, this is known as **rebasing**. Other tools may refer to this as **integration**. The point is to keep up with the **main** branch so that you're ready to merge your efforts when done.

If you're unfamiliar with this concept, it may sound a little petty to be so focused on keeping everything GREEN. But if you look below the surface, there is a lot of inherent value in this approach.

A main branch that the entire team ensures stays GREEN is always ready for release. Perhaps you have scheduled a release for the end of the month, but a critical update was just merged yesterday. If your manager comes to you and says the customer needs that feature immediately, it's no big deal to cut a release right now.

If you make that release at the end of the month, but there was an undetected bug, then it becomes easy to fashion a test case exposing it along with a patch, and issue another release two days later. That's because your confidence that the rest of the system is operational is much stronger than what we had back at the start of my career.

CI servers also have the power to do branch builds, verifying that you and your colleague's branches are operational as well.

This is the *real* point of continuous integration.

Not simply building every commit on every branch, but always working toward an *operational state of things*, be that on the **main** branch or any other relevant branches.

Of course this demands that you embrace thorough testing as discussed in ***Rule 6: Test What You Build***. If you don't have an automated test regime then your CI server will end up doing not much at all.

WARNING: I just mentioned the ability to let CI servers build contributions from community members. However, there is a certain amount of risk when you run build jobs from outside contributors. While we'd like to assume the best, it opens up the system to various attack vectors. What if someone creates a patch request that simply replaces the code with a script that takes supplied credentials and emails them to the attacker? This is why CI servers need to guard against such things. Sometimes it's easiest to only run branch builds from trusted sources like teammates.

Is buying into the concept of continuous integration all it takes to make your code ready for production?

No. There's more.

That's because you not only need a full armada of automated tests; you also need to consider running alternate builds.

For example, if your application supported both Hibernate 6.1 and Hibernate 6.2, it makes sense to run two different jobs, one for each.

In order to do that, does your build system even *allow* you to swap versions like that? If it doesn't, you need to first backup and ensure *that's* possible.

Maven has built-in profile support that makes it easy to change version numbers, property settings, and other things; Gradle has something similar.

Check out the following screenshot from Spring Data JPA's CI pipeline. It shows how many different jobs that team runs to ensure their code is ready for release:

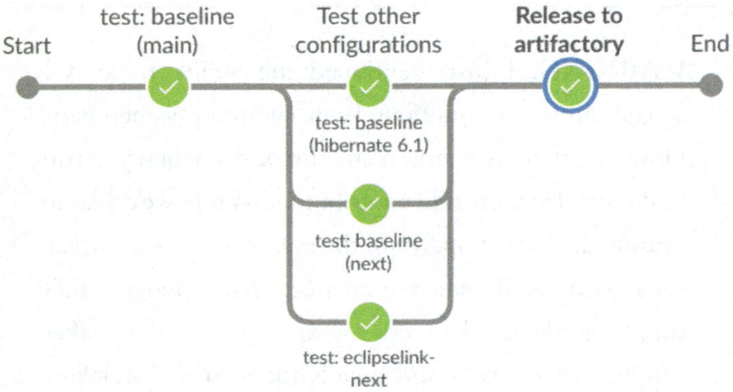

In this particular CI pipeline, the first phase has a single job, **test: baseline**. That job runs the main test suite. This is their baseline for being operational. It uses the minimal version of Java, the standard versions of libraries they interoperate with, and a basic level of 3rd party databases that they integrate with through Testcontainers. If it fails, then something is seriously broken, and so there's no need to run any more of the pipeline.

The second phase, **Test other configurations**, shows three different jobs that all run in parallel.

- The first job, **test: baseline (hibernate 6.1)**, verifies that everything runs on Hibernate 6.1 (default is Hibernate 6.2).
- The second job, **test: baseline (next)**, verifies everything works with Java 20 (default is Java 17).

- The third job, **test: eclipse link-next**, verifies EclipseLink 4.0.0 (default is EclipseLink 3.0.3).

The final phase, **Release to artifactory**, is where they publish a copy of the JAR for consumption by the community.

As you can see, this is a lot more than simply compiling and running the test suite.

So what does this have to do with ensuring your code is ready for production?

For starters, it's important to have that suite of automated tests. Yes, I'm going to keep hammering that point home!

But it's also important to ask "In what *other* scenarios will my library or application be used?" as well as "What does it take to build against more than one version of a separate library people will be using?"

Part of this drives the Spring team to have limits on how long they support a given version of their projects. They also have limits on how many versions of a 3rd party library they will support. Spring Data JPA **3.2.x** only supports Hibernate **6.2** and **6.1**. If someone needs Hibernate 5.x, they have to go back to Spring Data **2.7.x**.

This type of logistical planning is also vital to ensuring your code is ready for production. And it also allows you to put limits on how far you must go in testing support for things.

If one of your users opens a ticket asking you to add support for a feature not currently supported, you aren't obligated to take it on.

Focusing once more on **continuous integration**, it's important to highlight the fact that testing every change, not being surprised by changes in 3rd party libraries, and focusing on relevant branches always being operational and thus mergeable or releasable, generates a much stronger confidence as well as a better ability to roll out changes to production.

Another critical thing your code needs before being released to production is oversight.

Having fellow colleagues review your efforts and provide feedback is vital to ensuring code quality. The tactics used in prior decades often involved tedious tasks, sometimes including writing long MS Word documents, printing out reports, and then gathering everyone in a conference room to review documents as well as the code itself.

But with the rise of GitHub, code reviews are baked right into the platform so many use that to warehouse their code!

All you have to do is create a branch, make your changes, push that branch to GitHub, and then open a **pull request**. It's kind of like a ticket with the ability to make comments. But it offers the ability to look at the changes, and also to conduct a review by making comments on various lines.

Check out the example below:

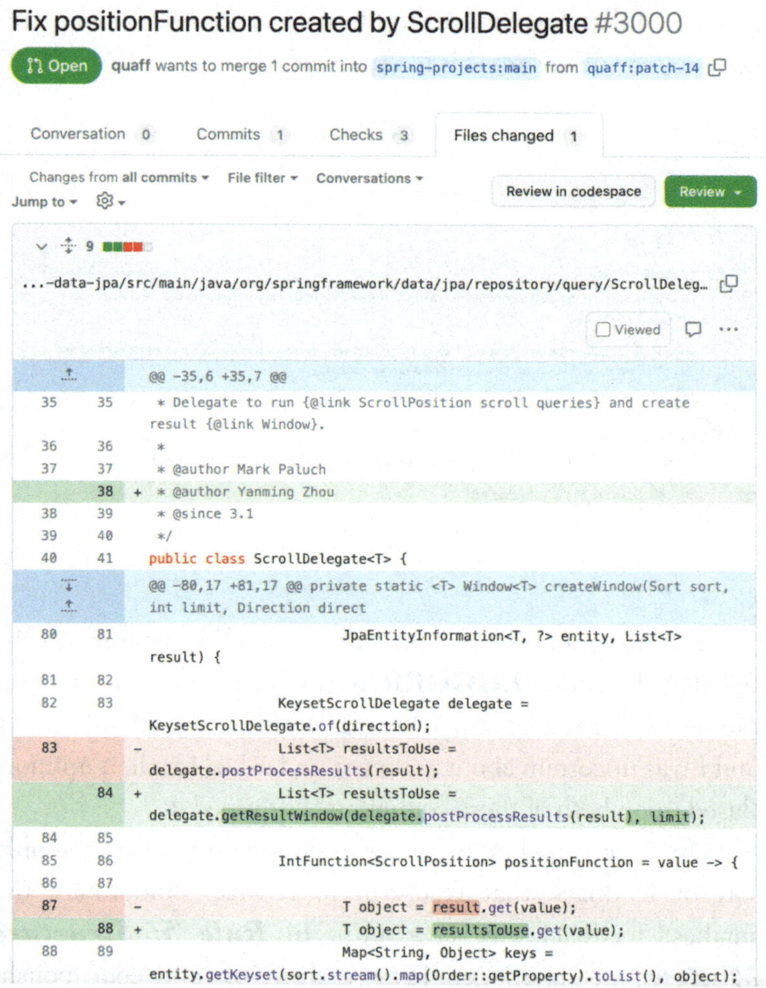

This screen capture from a pull request highlights the various changes submitted, with the lines and tokens highlighted (also note the "+" and "-" symbols on the left denoting the lines being added as well as removed).

This makes it super easy to skim and focus on what changed. What GitHub also includes is the ability for a teammate to provide a series of comments in the form of a review.

The following screenshot from a previously completed code review illustrates this feature perfectly:

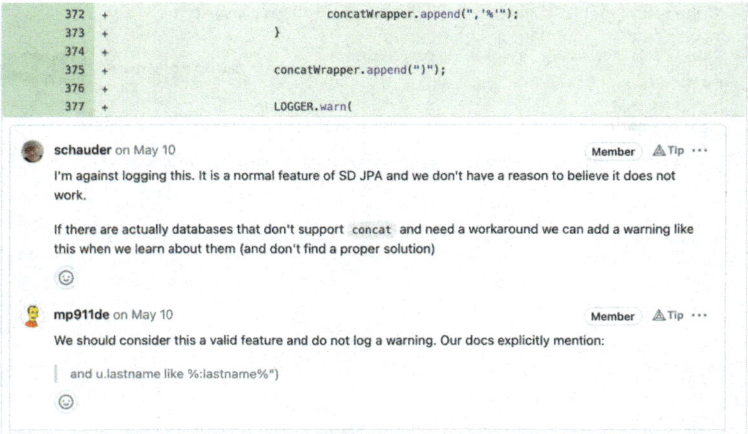

The first comment was provided by a fellow teammate who felt that the added **LOGGER.warn()** statement was unwarranted. To be transparent, this was a pull request I submitted, and I was uncertain about adding it, so I asked for their opinion. Based upon both of these comments, I dropped it.

Opinions on whether or not to do something isn't the only sort of feedback that is useful to receive. Sometimes, the smallest feedback can be useful. In ***Rule 5: Structure Matters***, we mentioned **juergenization** and code polishing. One of the most useful things *you* can do in the midst of a peer review requested by someone else is to point out spots where the pull request isn't upholding the team's standards.

Look at the following screen capture from a past pull request:

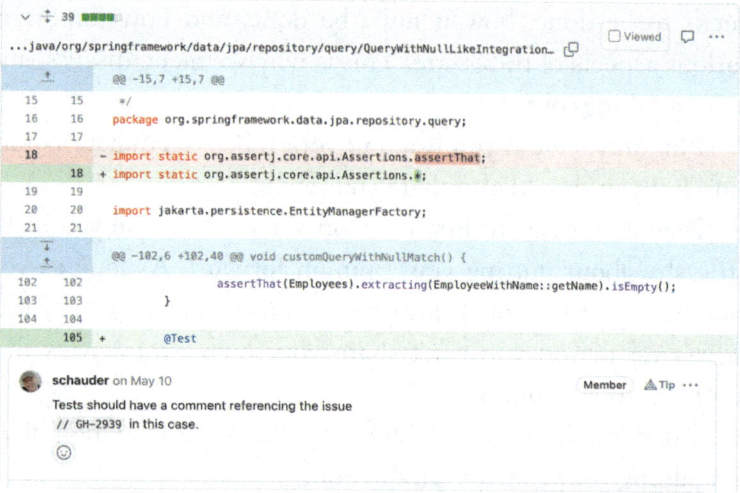

On the Spring Data projects, their standard is to document the source on every test case by citing the ticket number as a comment. It provides a quick way to go back to the issue tracker and understand the bigger picture that produced that test case.

In this comment, my teammate pointed out that I forgot to include such a reference to this new test case. There is nothing rude or antagonistic here. Simple constructive feedback.

Which leads us to an important aspect of providing feedback.

There is no reason to be suspicious or doubtful of a fellow developer's intentions. Instead, it's best to start from an assumption that our colleagues are trying to help us.

If you find this challenging, perhaps your team is overdue for some team building exercises? One of the most important things your team can do is build mutual trust and support. In fact, there are many resources with ideas on how to gather as a team.

It's important to be able to work together as a team. But people are sometimes confused by what that means.

Consider this. We all have differing opinions on what

needs to be done, how it must be done, and how important various aspects of things are. This is why we meet, discuss, and hammer things out.

Not only should you listen to your fellow teammate's opinions, but you should also share your own.

Maybe this is your first week on a new team, and you feel a little shy about putting your opinion forward. As long as you maintain a professional attitude and avoid pushing too hard, there's no reason you can't share your own experiences that inform your own opinion!

Good feedback from trusted teammates is a vital ingredient in building your code for production.

While demo apps are thrown together for quick presentations, production code is forged through steady refinement and polishing. Production code, unlike demo code, is developed for the long haul. Spending a little extra time (hours or days) can be worth it to avoid rushing hastily written and sloppily test code.

This is why it's perfectly valid to comment on small things (proper comments) as well as big things (the basis of the design is way off). There are, in fact, many areas that it's important we give teammates feedback, which we'll cover in the next section.

RULE 8: MANAGE YOUR TICKETS_

When we write code, we make mistakes. We can't anticipate all situations. And we often have to design new features.

This results in opening tickets.

Tickets are a constant drum beat of managing changes to code. While simple in concept, there are some valuable lessons we'll tackle here in ***Rule 8: Manage Your Tickets***.

THE FIRST STEP in any ticketing system is to agree on where the tickets are kept. We've already mentioned GitHub multiple times in previous chapters. GitHub's tickets, or ***issues***, provide a top-notch system. Any ticketing system provided by your source code repository will often have the most integrated features. And GitHub has no shortage of useful features.

NOTE: This book is in no way sponsored by GitHub. I have used multiple ticketing throughout my career, and

> GitHub issues seems to turn out to be one of the most effective systems I've ever seen.

With GitHub issues, it's handy to carve out some labels to track various tickets. Labels are metadata and give you the ability to assign more information to tickets. This may sound like a petty feature, but it's not. Labeling tickets can give you more at-a-glance ability to know the scope, importance, and area of impact for a given ticket. We'll talk more about what all that means, but first, let's explore some useful labeling options.

While GitHub's labeling system is quite wide-open, meaning you can do literally anything, here are some suggestions on what makes for effective labels starting with **type**. Here are all the **type** labels found in Spring Data JPA:

type: blocker	An issue that is blocking us from releasing
type: enhancement	A general enhancement
type: task	A general task
type: documentation	A documentation update
type: regression	A regression from a previous release
type: bug	A general bug
type: dependency-upgrade	A dependency upgrade

Each of these is a label (on the the left) with a common prefix (**type:**) and is configured with the same color. The implication is that a given ticket should be assigned one of these labels. You can read the descriptions on the right to better understand what each **type** means. This nicely lets you differentiate between bugs, enhancements, documentation updates, and critical blockers.

Another axis of ticket labeling is the **status** of the ticket, indicating where it is in the workflow of ticket management. Check out the following list:

status: duplicate	A duplicate of another issue
status: invalid	An issue that we don't feel is valid
status: declined	A suggestion or change that we don't feel we should currently apply
status: blocked	An issue that's blocked on an external project change
status: superseded	An issue that has been superseded by another
status: feedback-provided	Feedback has been provided
status: feedback-reminder	We've sent a reminder that we need additional information before we can continue
status: waiting-for-feedback	We need additional information before we can continue
status: ideal-for-contribution	An issue that a contributor can help us with
status: waiting-for-triage	An issue we've not yet triaged
status: pending-design-work	Needs design work before any code can be developed
status: on-hold	We cannot start working on this issue yet

Now you don't need *all* of these **status** labels, but hopefully you can at least see the difference between **status** and **type**. Again, these all have a common prefix in the label's text (**status:**) as well as a different, common color from the previous set of labels.

IMPORTANT: These labels all have both a color and text. The color is supposed to offer a quick recognition of the label's type, but some people can't see color the same as others. That's why it's also a good thing to embed the label's taxonomy into the text itself. A yellow label of status: on-hold becomes very clear from multiple angles what it really means.

A third axis to consider capturing as a batch of labels are the various *components* of your application. I haven't included Spring Data JPA's set of components, because they probably wouldn't align with your application. In lieu of that, here are some recommended labels to create for a typical web application:

- **in: web**
- **in: service**
- **in: persistence**
- **in: security**
- **in: messaging**

This is one way to structure a collection of labels using a layer-based architecture. If you are using something other than a layer-based structure, you need a different set of labels matching up with your app's components.

But no matter *what* structure you use, a key aspect is that they all have the same prefix (**in:**) and these labels should all have a common color between them. Each component-based label should clearly denote what it represents.

Perhaps this doesn't perfectly fit your app? Feel free to customize the list and add more labels while cutting out unnecessary ones. For example, if your app doesn't have messaging, no need for an **in: messaging** label.

Another thing is to make labels based on logical grouping. As I said, this is for a layer-based architecture. But if you have split up your application into various verticals, then it makes sense to have a set of labels that match them, e.g. **in: orders**, **in: invoicing**, **in: administration**, etc. And it's up to you whether you think the layers and the verticals should be the same color or different. However your application is split up, you should have corresponding ticket labels.

Now we seem to be spending quite a bit of effort on building various labeling mechanisms. Like ***Rule 5*** says, ***Structure Matters***. Investing a little upfront effort in designing a solid ticketing system will help keep order intact and avoid your system falling into disrepair and your ticketing system to become a drag on productivity instead of a tool to manage evolution of your system.

It's also important to point out that things evolve. Systems change.

It's okay, perhaps once a year, to go back and review all your labels. Maybe you picked up either a new layer or a new vertical of functionality. I give you permission to go into the system and add new labels.

It's also important to ***never delete a label*** unless it was a mistake.

Deleting labels because they are no longer relevant will actually remove them from old tickets. I wish that GitHub had the means to archive labels, but they don't. Instead, if you removed a certain component and need to do away with its corresponding label, edit that label. Change the text to something like **in: messaging [archived 10/21/2030]** and change its color, perhaps to some sort of flat gray to give it the appearance of being grayed out. That way, you can still hold on to the contextual information, but also be made aware that it's no longer relevant for new tickets.

Arranging all your labels to identify **which** components had what **type** of issue and what **status** of the process is the first step toward ticket management. The next section will talk about various tactics to successfully prioritize your work and organize your work schedule.

After spending a little bit of effort setting things up, it's time to start using it. And while spending a few hours setting up your ticketing system is valuable, getting into a rhythm of ticket management is what really makes a difference.

Plan out a time to look into new tickets. Either meet in person or on a Zoom call, once a week, in which you go through all the new tickets that have yet to be looked at. This is your **triage meeting**.

Just as a nurse will first check a patient for key symptoms at the doctor's office, your job is to check out initial tickets to see how critical they are.

Show-stopping bug? Mark it up!

Something already resolved? Link it to the relevant ticket and close it.

Something that needs more investigation? Decide who will dig in a little deeper and assign it to them. No need to assign its label during the meeting.

NOTE: It's possible to configure GitHub to apply default labels when tickets are first opened. This is a great place to configure your system to apply that **status: waiting-for-triage** label.

The focus of the triage meeting is to simply disposition new tickets and to also decide if they should be addressed immediately or in the future.

For the record, this is the process the Spring Data team used when I last worked with them. We would meet on the first day of every week. Pick a time that works for your team. What's key is finding a consistent pace to review incoming tickets. It's more important to identify who has the best means to assess a given ticket and assign it during the meeting rather than come up with a solution.

A subtle thing in the midst of all this is agreement between teammates on who covers what. For example, on Spring Data, we each had various parts of the portfolio that each of us was more skilled at. One of our team members tended to manage MongoDB issues, while another usually tackled Redis issues. I had introduced the ANTLR-based query parser for JPA, so any parser-specific tickets were given to me.

This is where it's good to let your various skill sets emerge as to who handles certain types of tickets.

Another thing is how to handle tickets no one wants.

These are the reasons to have a team leader. If you're not certain whether you should take on a ticket, perhaps because you feel that you don't have the skills, it's okay to voice that concern during the meeting. "I'd love to look at that, but I'm not sure my Redis skills are up to speed yet."

Another avenue to figuring out how to deal with a ticket is

to agree to take on a ticket, *but only* to assess it further. For example, during the triage meeting, you may not have enough information to decide if you can fix it. But you could agree to spend a couple days digging in and getting familiar with it. Then plan to come back at the next triage meeting with more information. In such a situation it's okay to let the ticket get assigned to you, but to retain its **status: waiting-for-triage** label.

And what about new feature requests from your customer or community?

Sometimes before moving forward on a ticket, the team needs to decide if a particular proposal is even agreed upon. It's not uncommon for someone to ask for a new feature that would take a lot of effort, make things more complicated, or serve only a small niche of users. Or some combination of those factors.

It's also okay to say no to such requests.

Let me repeat that: **it's okay to say no to some requests**.

At the end of the day, it's you that will be maintaining things. It's you that will fix its bugs. And it's you that has to explain the benefit of this new feature to others that had nothing to do with its design.

We once added a new feature to Spring Data JPA called the Query Rewriter. It was a callback right before a query was handed over to the entity manager that allowed the developer to apply any custom alteration to the query. It was my idea, my implementation, and became my thing to explain to a fellow developer advocate during an upcoming live stream I was about to join.

And that's when it hit me—my Query Rewriter feature that I had worked hard to implement in a nicely coded fashion was short on value. It was really a stopgap to deal with the fact that we didn't have a fully fledged JPA query parser.

When our developer advocate asked in advance what were some scenarios to use this Query Rewriter, I had to level with him—none.

None that I could think of.

And that's when I knew we had taken on a less than ideal feature.

So if possible, try to avoid agreeing to implement something that isn't well-defined, doesn't have a good example use case, or doesn't have a clear need among your users or community members.

These are all valid things to discuss during a triage meeting.

After conducting your triage meeting, when you have your updated list of tickets assigned to you, it's time to start working on them!

Hopefully, you have a better sense of what's expected of you. Team expectations are the first step toward having a prioritized list of things to do.

At this point, I'd like to point out GitHub's project dashboard feature:

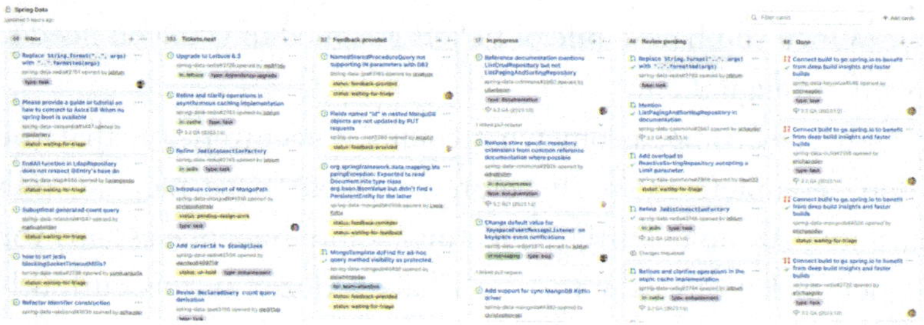

In this example, again from the Spring Data team, we had the following columns:

- **Inbox** - new tickets that haven't been

dispositioned automatically have the **status: waiting-for-triage** label applied
- **Tickets.next** - tickets that are properly assigned and labeled, waiting to be carried out
- **Feedback provided** - tickets where the user has provided requested details
- **In progress** - tickets being worked
- **Review pending** - tickets completed and awaiting review
- **Done** - tickets completed and closed

This ticket board can help give you a task list to work on. It's quite easy on GitHub's ticket project dashboard to click on your avatar and see only the ones assigned to yourself. You can also grab the URL for this dashboard and drag it to your browser's **Bookmarks** bar at the top, making it easy to go back to this board.

And this is where the rubber hits the road.

We've talked a lot about how to label tickets, how to disposition and assign them during a triage meeting. But at a certain point, it's time to dig in and code.

Once you have a pile of tickets assigned to you, you need to work through them.

Fair warning, a somewhat common occurrence is that you don't complete them all before the next triage meeting rolls around. At the next triage meeting, before someone assigns you a new ticket, it's okay to say, "I have a dozen tickets currently open. Is this ticket more important?"

If your team or manager says, "yes," then you can accept the ticket. If not, they might assign it to someone else. Or they may leave it unassigned for now.

Did you notice the way I suggested you bring it up?

Don't say, "I can't handle another ticket!" and begin to moan.

Instead, describe your situation without emotion, and indicate you need to know which ticket gets top priority. Sometimes it's obvious, other times it requires a manager who is plugged into the other teams and the overall schedule to let you know which tickets need to move ahead of others.

There is no requirement to "finish all the tickets." In fact, there will *always* be more tickets. Our role is to simply work the most important ones.

To do that, you must stay in frequent communication with your team leader and manager, so they are aware of your workload, what you've completed, and ensure you are getting direction on what's most important.

It's up to our leaders to step in when appropriate and tell us to *not* accept a new ticket or to indicate, "we need help and are beginning to interview candidates."

There is a wrinkle to GitHub's project dashboard—it may not show you *everything* that is on your plate. For example I actually managed a separate Spring project when I was on the Spring team and it wasn't part of Spring Data's ticket dashboard.

I had to go to a different place and check out that backlog of tickets from time to time. There was also a system where our company hosted training courses we had to complete on occasion. There were channels where members of our community could raise questions. And there was a Google Doc I used to collaborate with a fellow teammate over a conference talk.

All of these various locations demand our attention and as a pro coder, it's up to each and every one of us to monitor them properly. This takes thought and consideration. The more things we have to manage, the harder it can be.

It's one of the reasons I started a daily work log. I've

mentioned this earlier, but to recap, I have a Google Doc that is just for me. I use it to journal my activities every day. If I so much as look at a ticket, I write it down along with a hasty description of what I did. Even if I do nothing, I write it down. This has encouraged me to be more meaningful with my time.

It also gives me instant feedback, because at a glance I can see all the various modules I've looked at in the past week. If one of my modules isn't appearing in my journal, it's a signal that I may be neglecting it.

This daily log also helps with the fact that after a week, there is no way I can remember everything I did.

When my manager wants a quick summary of what I've worked on over the past month or quarter, I have this nice record of action showing everything I've done. I can extract the details he needs.

It also has another vital benefit—it provides me with positive encouragement about my work.

It's very easy to forget all the positives and instead focus on a small handful of negatives. Having this catalog of work always gives me a warm fuzzy when I log yet another task. Sometimes I pause my work tasks and simply scroll for a few pages, realizing how much I've accomplished even in just the past month.

It's amazing what a single minute of pure scrolling can do to lift your spirits every few days.

Oftentimes, that sort of quick review of my work absolutely clobbers any negative feelings that start creeping in.

It's also good to go into your journal at the end of your official work day and ensure you logged everything. And then... close it all up. Sign off of your desktop station or close up the laptop. Either way, a critical part of staying organized is to stop working.

Instead, leave your office and stop. This is especially important if you're working from home.

It's easy to sometimes drift into an extra hour of work. You're not as likely to work extra hours if you have a family. But back when I was single, there were times I would stay late.

Don't get me wrong. Every now and then, it's okay to put in a little extra effort. But that should be the exception and not the norm.

And it's important to remember that when you start working lots of extra hours, your effectiveness becomes diminished.

There are some studies showing that in a given work day, if you manage to get a solid four hours of work completed, then you've done great. In fact, a couple of those four-hour work days in a given week should be celebrated and not bemoaned.

Considering most of us work on an eight-hour work day cycle, this may sound awkward. Perhaps you're assuming that anything less than a solid eight hours in that allotted time is inadequate?

It's not.

In fact, it's vital that you take breaks.

The more you take breaks, gather your strength, and refresh your mind, the better you will get at achieving those four solid hours of coding.

This is why pivoting between full-on coding and updating your journal with things you've done is a good way to work toward that. Taking the time to update the status of tickets you've worked is another task that needs to happen and can also serve your need to give your brain a break.

Or pause and watch a YouTube video. Or read a chapter of a book.

We already went into the importance of taking breaks in ***Rule 2: Building a Workplace***, so let's not dig any further.

Just know that it's hard to be the best version of you, if you don't take them.

However, managing your tickets along with logging your efforts isn't the only thing that will be expected of you. Interacting with colleagues as you knock out tickets is a critical task.

ONE OF THE immediate things we need as pro coders is feedback. Pure, honest, encouraging (not discouraging) feedback.

Of course, I speak of code reviews.

Early in my career, a code review involved printing out multiple copies and delivering them to all relevant parties. I had to strap on my sneakers and physically walk printouts of code and/or design documents to the project lead, my manager, the head of security engineering, and a couple other people.

Yawn!

Today, the game has changed. You can fork a repository, cut a new branch, make your changes, push it to your clone, and submit a pull request. Apart from the part where you're actually making changes, the rest can be done in minutes (not days).

Make your personal request of who you want to review it and sit back and wait for the comments to roll in.

The process is almost effortless. Well...the *tooling* is almost effortless.

I've had code reviews that spanned several days. Some that involved nothing but minor tweaks, like fixing a misspelled word. And I've had code reviews that involved junking the entire effort and starting over.

So let's first talk about the mechanics that flew by a couple paragraphs back. People like to think that Git as a version control system is quite powerful, but the honest truth is that

GitHub the website is what (in my humble opinion) truly made Git take pole position in the race for better source code management.

If you're already familiar with GitHub mechanics, you can skip forward a few paragraphs to the code review part. If not, buckle up.

GitHub is a decentralized version control system. That means that when you **clone** it to your local machine, you're getting *all* the history. You can make changes, switch branches, commit your changes, and fashion new branches, all without talking to the server. You can do the bulk of your work proverbially on a plane flight with no WiFi.

There are many commands for git's CLI (Command Line Interface) to use inside a terminal, but you only need a fistful to effect change. **git clone <repo url>** is the one that will grab a repository on the website, and put a copy of it on your machine, in the folder you're currently in.

Now an important consequence to understand is because it's detached from the server, you won't automatically get updates. If a colleague pushes changes to the **main** branch on GitHub, you won't see them without deliberate action.

If you are currently looking at the **main** branch on your machine, and you want to catch up, type **git pull**. This literally pulls the updates for that branch from GitHub into your clone.

The flip side of all this is that if you make changes on your clone, they don't automatically update the remote version on GitHub.

When you make changes on the **main** branch on your machine and want to push them to the server, the counterpart command is **git push**. This will, as you probably guessed, push your changes to the same branch on GitHub.

Never fear!

If your teammate pushed changes to the **main** branch on GitHub after you made your clone, and you never did a **git pull**, but committed changes anyway, GitHub will stop your attempted **git push** dead in its tracks. It will put up a warning message that there is a mismatch you need to resolve.

The simplest change you can make is to pull your teammate's changes from GitHub and then **rebase** your changes against that. To do that, type **git pull —rebase** (note the double dash). This will take your local changes on **main** and set them to the side, update your copy of **main** and then recalculate your changes against the update.

If the thought of making edits right on the **main** branch is making the hairs on the back of your neck rise in fear, don't worry! A lot of teams use branches as a way to manage work. Just type **git branch my_new_branch**, and git will create a local branch based on the branch you are currently pointed at. You can then **git checkout my_new_branch** and start making changes.

Make all the changes you want on your branch. When it's time to commit them, type **git add .** (don't forget that period!) to add all the local changes. Follow it with **git commit**. You'll be prompted to enter a commit message. Fill out all the details you want (See ***Rule 5: Structure Matters*** for tips on what to put in your git commit messages), and exit the editor. You can then **git push origin my_new_branch**, and your changes are pushed to your own side branch on GitHub.

Voilà!

At this point, it's easy to work toward merging your changes to **main**. You start by typing **git checkout main** followed by **git pull** to pick up the latest changes. This way, whatever you missed while working on your local branch will be pulled down to your machine. Next, type **git checkout**

my_new_branch followed by **git rebase main** (Search the internet for "interactive rebase" for more options). This will update your branch to the latest changes that have happened on the "main" branch, while keeping your work on your branch.

From here on, I'd really recommend finding an introductory book on Git to dive a little deeper. What we just covered is the basics of using Git. It's neither complete nor extensive. It's but a taste of a simple workflow.

With that fast introduction to Git mechanics, let's pivot to something a little deeper...how to conduct a proper code review using GitHub.

Assuming you have pushed your branch to GitHub as mentioned previously, you can then visit GitHub, navigate to your branch. Toward the top, you should expect to see something like this:

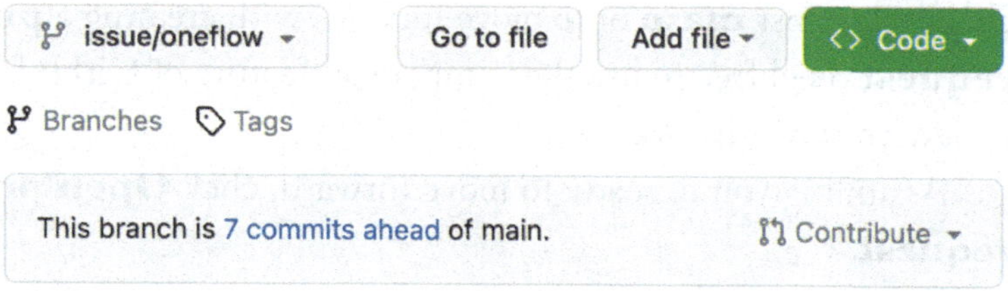

The top-left box is your branch. Confirm that it's showing your branch.

The status bar below it should show your branch's status relative to **main** (assuming your branch is based off of **main**). Notice how it has a hyperlink regarding the commits you have created?.

To create a pull request and get feedback, click on **Contribute** in the bottom-right corner. You should see something like this pop-up:

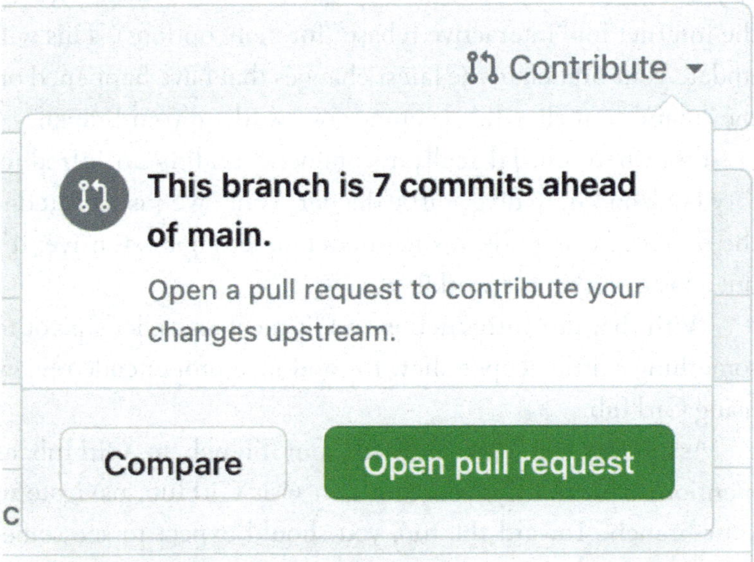

The pop-up will offer the option to either **Compare** your changes against **main** or to move forward with creating a **pull request**. Feel free to use the comparison feature of GitHub to review your handiwork.

Assuming you're ready to move forward, click **Open pull request**.

TIP: In case you didn't know, a **pull request** is when you are asking the project to *pull* your changes from your branch to the **main** branch.

You should see a form like this:

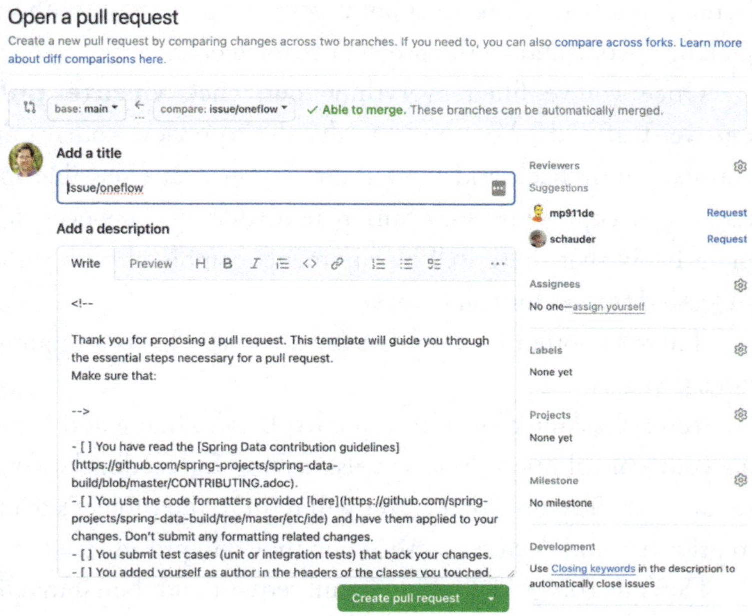

At the top you can see that this is a form for submitting a pull request. The status bar shows that the intent is to merge from the branch **issue/oneflow** to the branch **main**. Below that is the title. By default, GitHub is pasting in the name of your branch. It's better to replace that with a more descriptive title.

Following the title is a box to fill out the description of your pull request. This particular project has a pre-built template that the project maintainers wish you to fill out. This is markdown, so simply adding an "X" to each checkbox item will signal you have taken all the steps needed to prep this pull request.

Finally, on the right hand side is where you can pick specific people and request that they review your pull request. If it's a project you have worked on before, you may have insight into who to contact. But if you've never made a contribution, you can instead submit your pull request without picking anyone and let the project's team decide.

Once you've filled everything out, click **Create pull request**. If and when the team decides to review your work, you can come back and view their response. It's possible for them to accept your contribution immediately. However, it's more likely that there will be comments combined with some suggested tweaks for you to make.

There's another subtle point to be made. This is a request, not a demand.

You are asking others to stop what they're doing and focus on your contribution. You are also asking for feedback. Your request just like any other ticket will require the project's team to prioritize and decide exactly when they will process it.

Don't do things like submit a pull request and then immediately write a post on Stack Overflow asking the same question. That's called double posting and it's quite rude. Instead, focus on other tasks you have on your plate. There is no guarantee the team is going to pick up and look at your pull request this day, this hour, or even this week.

Now if this is a pull request submitted to the very same team you are working on, you can alert your teammates through whatever Instant Messaging tool you use that you have opened a PR and would like some feedback. Even include a link.

And then you must let it go.

Don't forget to pencil that into your daily work log that we spoke about back in ***Rule 1: Curate a Professional Persona*** to help battle imposter syndrome!

This is how to get feedback on an existing branch of code. But that's not the only area you must interact with regarding community members and colleagues. The next section will show more ways to improve your means of interacting with teammates.

What does it mean to collaborate with your team?

If you're a junior developer, it's unlikely you'll be handed an assignment to carry out all by yourself.

It's quite common to be paired up with someone else, such as a fellow coder of similar experience while being directed by a senior engineer, or to be working directly with that senior engineer. And to make those hours of working together count, there are some strategies you need to learn.

We've already talked about triaging and organizing tickets. That's the first step. Tickets are where we get our tasking. We use them to identify what systems are affected as well as how critical the situation is. And we assign the tickets to understand who is responsible for resolving them.

The next step is to work the ticket.

First, read the ticket to understand the problem. If the one that opened the ticket has provided little in detail, you can always ask for additional feedback. On the Spring Data team, if we needed more detail about a given ticket's situation, we'd add a comment indicating what we need, change the ticket's status to **status: waiting-for-feedback**, and then take it off our project dashboard. We had the GitHub project dashboard configured that when a comment is added, the status switches to **status: feedback-provided**, and the ticket is put back on our project dashboard.

Whether they provided us all the details we needed up front, or if they provided it later at our request, the next step is to dig in and try to reproduce the problem ourselves. This can involve simply running a given test case. Or they may have provided a sample application that exposes the issue.

Spring Boot made the concept of providing a **reproducer** quite easy, that is a tiny self-contained sample application that demonstrates the problem. And GitHub makes it easy to create new repositories with this **reproducer**. Then the user need only include a link to their newly-created repository in the ticket.

However the problem is shared with you, you may want to then consider capturing the problem into an automated test. This is how you move further into really understanding the problem. First, you reproduce it the way they imagined it. Next, you try to reproduce the same problem within the confines of your existing automated test machinery.

And sometimes capturing an automated test case with such an issue can be tricky. If you are somewhat new to automated testing, this is a step that may feel like it's taking longer than you'd like. But trust me, it's good to work through it, even if it takes hours or days to craft that test case.

As we discussed in ***Rule 6: Test What You Build***, building one or more automated tests that expose the problem will then ensure that when it's fixed, the problem doesn't creep back in again.

Another aspect of team collaboration is recognizing when you need help.

Perhaps the system the ticket was opened against is one you're relatively new to? It's okay to reach out, either through email or your team's instant messaging service (or simply walk over to your teammate's office), and ask for help.

One way to get help about a ticket you're trying to tackle is to carve out some time to simply talk about it. This can be face-to-face if you work in the same office, or it can be a Zoom chat. This could also be to a pair programming session, where one of you does the coding and shares the screen while you both work through it.

In these scenarios, it's okay to ask questions. You can always say, "I don't understand that. Could you explain it a bit more, please?" Or perhaps, "Why are you doing it that way? I'm not sure I see the reason for that."

Remember, you are both professional coders, and none of us have all the answers. This type of back and forth is healthy and only serves to increase the knowledge between each other.

We also mentioned in ***Rule 7: Build Things with Production in Mind*** that when reviewing a pull request linked to a ticket, it's okay to point out faults in the code and failures to adhere to coding standards. But it's also just as acceptable to ask questions like "I don't understand why this block of code is here."

Or you can simply ping them and ask for a Zoom chat to go over the pull request with the developer, share your screen, and communicate in that fashion. Remember, your teammates are trying to help you, not criticize you in a negative way.

Thus, when someone reaches out, try to afford them the same measure of respect.

Another vital aspect of collaborating effectively as a team is to not "keep it all to yourself."

I once picked up an application from another member of our team at a previous company. The person who had worked on this app had avoided tapping other teammates to review his work.

He had clearly sought out answers to things on his own and didn't have others inspect his handy work. And some of the coding patterns were quite atrocious. I already alluded to this situation back in ***Rule 4: Know What To Do***.

Simply put, this developer did NOT know what to do. They had tried to *not* disturb anyone and had effectively shut out any one else from pointing out places they could have made improvements.

Often we are given responsibilities to get things completed.

But it's okay to pause and ask for a fresh pair of eyes that aren't as deep in the code to tell us if we're even on the right track.

This, too, is a way to ensure effective collaboration with your team. Invite them in and let them see your efforts when the time is right. In a sense, the final code needs to have a smooth look about it such that you can't look at one section and say, "That's Alice's code. I can tell. And that over there is Bob's!"

Our code should have a consistent look-and-feel and design that transcends us, thus serving the end users. The sort of the thing that can only be achieved when we correspond with our fellow teammates.

There are also times when we simply need a decision agreed upon by the team. One such example was how the Spring Data team had used Project Lombok for years. At first, we used it everywhere, in both code released to the outside world as well our test code, that while open source, wasn't part of any published artifacts.

For various reasons, we decided to switch to *only* using Project Lombok for tests. We removed it from the public eye, but still wanted to leverage some of its unique properties to help maintain test scenarios.

When we needed to start testing against Java 21, we found out that Project Lombok taps into private APIs of the JDK that were no longer available in Java 21 and thus became a blocker.

This is the type of decision that extends beyond simple coding. Project Lombok had provided much value, but it had reached a point where we could no longer count on it to make releases fast enough. We had to either drop it or risk having our entire release schedule slide to fit Project Lombok's release schedule.

As a team, we chose to drop it.

This is the type of thing that can be covered during your

weekly triage meeting. Or if you have another periodic team call, you could discuss it there.

Meetings are the place to share your opinions and listen to others. And good communication that is always flowing is the real key to effective collaboration.

So what do you do if your meetings are *not* resulting in a solid flow of communication? Read the next chapter, ***Chapter 9: Hands Meetings Like a Pro*** and find out!

RULE 9: HANDLE MEETINGS LIKE A PRO_

Despite all the fun we have writing code, it's not possible to maintain a system of any size without having meetings.

Ugh! Not meetings!

If you are already dreading the concept of having meetings, then gear up. We are going to delve into many useful tactics to make meetings more effective. But before you can make a meeting worth the time...we have to discover *what* these meetings are.

The fundamental reason to have a meeting is to communicate with your fellow teammates. That can include discussing upcoming features, triaging tickets, reviewing finished work, presenting technologies, having training sessions, post-mortem analyses, and more.

There are several types of meetings, but the simplest one that has only been around for over a decade is the **standup meeting**.

Standup meetings were originally designed to be quick and short. We're talking five minutes at the most. No, not five minutes given to each person attending the meeting. Five minutes ***for the entire meeting***.

A standup meeting got its name because everyone attending would be standing up. No reason to sit down for such a short meeting, right? The point at a standup meeting is to say, "I'm working on a new query parser for Spring Data JPA!" Everyone else nods, having heard what you said. And then the next person gives their daily status.

It's also the chance for you to say, "I tried to build that new whatchamacallit yesterday, but couldn't get it working. Can someone pair with me for a couple hours?"

There's no need to discuss all the details at the meeting. Instead, you're there to share what you're doing and potentially what you need help with. With a team of a dozen people, each spending thirty seconds at the most, the meeting is wrapped up fast.

Standup meetings that involve a dozen people or less and run for no more than five minutes can be done on a daily basis.

There is also another rule you can choose to adopt. During standup meetings, ***only the people writing the code are allowed to speak***. Management can stop by and listen in, but they aren't supposed to be active members. The focus of this meeting is to keep the coding team up-to-date with what everyone is doing. This is also most effective when you're talking about a relatively tiny group of developers.

If you have fifty people on your engineering team, then a fifty-person standup meeting isn't very effective. However, it's not uncommon for such a large team to be divided up between various apps. Perhaps the whole team is managing five apps and each app has ten people assigned to it. Each of those sub-teams could decide to host a daily standup meeting so that everyone can keep up with what's happening.

Now I mentioned that standup meetings started as a meeting where literally everyone was standing. But it's possible to have such a short, daily meeting virtually through Zoom (or whatever your video chat tool is). It's *more* important to keep everyone up-to-date on what everyone is doing than to physically be standing.

Another meeting we mentioned in ***Rule 8: Manage Your Tickets*** is a **triage meeting**.

We went into this a little in the previous chapter, but it's worth repeating. A triage meeting is where your team can sift through new tickets, ensuring they are properly categorized into your system and assigned to the right teammate. When I was on the Spring Data team, we had a weekly thirty-minute triage meeting, and it was quite effective.

Before we started working tickets, we had to ensure they had enough information. This involved leaving a comment on the ticket asking for more feedback from the person who opened the ticket and taking it off the board until they responded.

A ticket with barely any details can be marked up with "Can you please describe your scenario in more detail?" Or perhaps "Can you create a reproducer using Spring Boot and Testcontainers to recreate the bug you are seeing?" With GitHub's ability to create canned ticket comments, it's easy for you to capture this language and mark up new tickets quickly.

During your triage meeting, try to assign each ticket based on who is most likely to know how to proceed. I would usually get new tickets for Spring Data JPA, while our lead for Spring Data MongoDB would get the MongoDB ones. That is also easy to do during a triage meeting.

It's also possible to set things up with a work queue.

Anyone that is free can grab the next one and start working on it. You could even make a special ticket label to indicate **status: ready to be worked**. You pick up the ticket, and change its status label to **status: in progress**.

There are also **planning meetings** where you meet in a conference room (or again, virtually in a Zoom chat) and go over all the features you are planning to implement. These meetings don't have to occur as often. Once, maybe twice, a year. The point is to get agreement from your teammates on everything planned for the next release. At least initially.

This type of coordination is good because it gives you an overarching plan on what to accomplish.

Early in my career, the team I worked on crafted Microsoft Word documents for each phase of our release process, printed them out, and had them delivered to various stakeholders.

In this day and age, it's much easier to create a Google Doc and build your next major release's list of target features in a more collaborative way. You can go ahead and share the link to your document with your teammates, either through email or whatever instant messaging tool you use. Send out a link, invite comments, and perhaps in the next two to four weeks, schedule a team call to go through the whole thing. A one-hour meeting should do the trick.

By communicating ahead of the meeting and investing time leading up to the meeting, you aren't spending all time *during* the meeting going through things for the first time.

Which brings up a *vital* piece of ***meeting etiquette***.

When you are invited to a meeting, you must carve out time to review any relevant information **in advance**. Going into a meeting and expecting the presenter to present everything can be seen as spoon-feeding, a rude tactic.

If material has been shared in advance, find time to read through it. Comment on it. Pose questions. This is part of being a pro coder. Don't forget to log this effort in your daily work log!

Now there are exceptions.

For example, in early 2023, I was given the task of putting together a proposal for dealing with JPA queries that were becoming too complex to handle, and I was asked to present it at our face-to-face meeting in Boston in a couple months.

I knew everyone was prepping for that week, and I had a bit of a surprise, so I didn't leak the details *on purpose*. But this was different than having a detailed design document. Instead, I had a Keynote deck I had carefully put together along with a demo project showing off a drafted ANTLR-based JPA query parser.

Not all your meetings should be a presentation. In fact, very few, since decks take a long time to put properly together.

Basically, try to determine in advance what you are seeking during the meeting and build your content appropriately. If you need agreement on several complex topics, then a high-level design document where you ask for people to review it in advance and then make choices during the meeting will alert everyone to your agenda.

I mentioned **standup meetings** earlier. There are also ***standing* meetings**. This is a meeting that happens at the same time and place, such as the same conference room every week (or the same Zoom chat).

On the Spring Data team, we had a weekly standing meeting where we would meet not just with each other, but also with the leads of the community projects that are outside the company, e.g. the lead developers for Spring Data Neo4j. At this weekly standing meeting, each participant was expected to cover the various things they had covered in the past week and indicate what they were planning to do over the next week. Our team leader advised everyone on when the next release would be, so everyone was aware of what we were trying to fit into the next release. And sometimes we discussed other topics that spanned multiple modules.

Because we had a weekly status meeting as well as a weekly triage meeting with the same people, we simply had them back-to-back on the same day (Monday). It allowed us to cover a lot of logistical planning every week and kept the whole team abreast of issues as well as progress.

But there are other meetings that you may not need that frequently.

There were also monthly meetings that involved all the Spring leads hosted by the Spring team leader. This was usually a bigger meeting and because of that, it wasn't as interactive. It tended to involve the Spring team leader sharing critical stuff that was on the radar for the next quarter. Sometimes one of the leads had some material to present that everyone else needed to be aware of.

All of these meetings required a bit of meeting etiquette. Whether you're in a face-to-face meeting or on Zoom, it's best to listen to others and try to be respectful. It's certainly *not* the place for heated arguments.

If you're new to all this, it may be more effective during the first few meetings to simply listen to the more experienced developers and gauge how they are interacting. Try to learn from them.

Don't worry—if you didn't get a chance to ask your question, you can always ping someone afterward. And if you accidentally interrupt someone, apologize.

If you're uncertain about an upcoming meeting or what role you are supposed to play, you can always ask the person organizing it, "What do you need from me for your meeting?" Your colleague will be happy to make things clear.

And anytime you're uncertain about when to speak up

during any sort of meeting, you can always have a 1-on-1 with your manager to get clarification. This is what your manager is there to do, help you learn the ropes and do your job.

Now if it's a little fuzzy about when you're supposed to talk and when you're supposed to listen, the next section should help make that more clear.

ANOTHER PART OF ***RULE 9: Handle Meetings Like a Pro*** is the fact that there are always circumstances where we are asked to contribute, and other times when we are asked to listen.

And I have a confession...this is something I have struggled with for a long time. I've always been prone to speak. Learning to listen has been *tough.*

Let's dig into an example to better elaborate what I'm talking about.

I've mentioned that weekly Spring Data team status meeting I used to attend. That was a meeting where each member was asked to describe what they did over the past week and what they were planning to do for the next week.

That is a situation where we are clearly being asked to contribute.

But our team also had monthly status meetings that could easily have 50+ people attending.

Such a large meeting, unless you're the one running it, is more about listening and absorbing things. If everyone asked a question, the meeting planner probably wouldn't have time to cover the material they put together. In truth, they may only have twenty minutes of planned material to go over and want to leave room for questions.

That is the situation where you need to be more careful.

If you have a question, try to get to the point quickly. Don't get stuck on long backstories. Instead, simply pose your question and then listen for the answer. If you're not sure about the answer, you can always ask about following up afterward.

Let's aim to respect everyone's time.

While that last sentence may sound like it only applies if you're attending someone else's meeting, it *also* applies if you're hosting a meeting.

If you're the meeting organizer there are some critical things *you* must do. This most certainly includes starting the meeting on time.

In general, I suggest that you respect the people that have shown up on time and start promptly. It's okay to indicate, "I think we'll start in one minute." But I'm a *big* fan of starting on time, on the dot. If someone is late, it can be seen as rude to those who joined on time if you delay it for those who didn't.

At the same time, you should end the meeting on time as well. Remember, you can always schedule a follow-up meeting. People have carved out time to attend your meeting, but may have other obligations right afterward. So don't force them to leave your meeting to make it to another. Watch the clock and call it quits if you hit the mark.

One last piece of advice regarding meetings and respect is the fact that tools like Zoom afford us the ability to record meetings. Sometimes for things like a weekly status meeting, not everyone can make it. We sometimes have personal obligations that can cause us to get to work a couple hours late. Having a recording of a status meeting is quite nice, allowing us to catch up.

But other meetings, ones where personal opinions are being sought, perhaps should *not* be recorded. People sometimes don't want to be put "on record" with opinions.

If you are hosting your very first meeting and aren't sure

whether or not to record it (if done virtually), you can ask your manager's or team leader's opinion.

But that's not the only thing we need when it comes to attending or conducting meetings. The next section covers some critical things any team needs to function well.

THE THING that makes it possible to have collaborative meetings, to give and receive feedback, and to work together to build systems is having a **strong team dynamic**.

You should strive to build trust among each other. Many of the things I've talked about in this chapter cannot be accomplished if your team's atmosphere is toxic.

What does it take to build such a strong foundation?

Trust. Plain and simple trust.

When you show a colleague a piece of code and they give you feedback, don't assume the worst.

Instead, assume that they are trying to give you constructive information to help you do better. Even if it feels a little rough at first.

Now I'm not asking you to ignore your feelings. It's possible to accept feedback and still express your emotions. Instead of responding with, "You're wrong! It isn't *that* bad!?" Try something more like, "Wow. I didn't realize that. I'm still new at this and trying to get the hang of code reviews, but I'll try to consider what you said."

The first response is combative and puts you and your teammate in an adversarial situation. You are basically demanding that they prove their position.

The second response lets you express that you are trying to process their feedback and it has affected you, but you are working through it. It also doesn't directly challenge them, but instead gives you a chance to decide, perhaps after you've cooled off, that the feedback was indeed correct.

Being open to feedback, being willing to think about and consider inputs from others, and being able to say, "Wow, I was wrong on that" is a great way to show respect to your teammates.

It's also important that if you feel strongly about something, that you say so.

Your team leader, project lead, or manager often don't have all the information. They are counting on you to do your part. There may be limits on what can be done due to technical and logistical limits. But they need to know *your* perspective.

Don't assume everything is about accepting the word of others.

As an example, my team had thought for a couple years about whether or not we needed a JPA parser. Every time the topic came up, the answer was, "That's a lot of work. I'm not sure there is enough pay off to go down that road."bnf

But when our leader asked me to come up with a proposal to deal with an increasing backlog of tickets concerning JPQL and HQL queries, I carved out a couple weeks to investigate deeper and learned how to use ANTLR, the parser generator toolkit. I found JPA's official parser in BNF (Backus-Naur form) and put together a demo proving its feasibility.

I had actually worked in building parsers in college, both as required undergrad coursework and later as a Graduate Teaching Assistant when I was earning my Master's degree. I had also built a custom parser for two prior work assignments.

I put together a presentation to pitch my idea.

I knew the risks involved and even fashioned a slide of some of the biggest questions I anticipated (along with commensurate answers).

By the end of my presentation, the two people I had to

convince, our team leader and the prior lead of Spring Data JPA, simply replied that I had answered all their questions.

You see, there are times when you are needed and your unique set of experiences is vital to the team's progress. The secret is doing it with a respectful disposition.

So what if you happen to have joined a team that seems to be toxic?

First of all, how can you even tell? Do colleagues speak down to you? Do they dismiss any ideas you have because you're new or a rookie?

Are you able to take issues to your manager and get help?

If one member of your team makes snide comments against you or others, and you feel too intimidated to speak during meetings, that could also be a sign that you're in a less than optimal environment.

One of the first things to consider is in fact going to your immediate manager and sharing your concerns. You can schedule a private meeting between the two of you, and open things up by saying, "Is this meeting anonymous?" You *should* be able to express your feelings in private.

If that doesn't work, and your manager breaches your request for anonymity or fails to listen to your feedback, then the odds of getting further assistance may be quite slim.

That is also a strong signal that things probably won't improve.

If so, what's the best path to take? You can try escalating things to Human Resources (HR) and working through it. I have never personally been through such a situation, so I can't tell you if this is a good idea of not. The alternative to fixing things on the inside may be to find a position with another shop.

Only you can determine if your situation warrants leaving.

If you plan to leave, don't be hasty about it. If you leave this company and *then* try to find a new position, you may find yourself backed into a corner paying your bills, etc. If you have decided to leave, avoid burning any bridges. Instead, try and work out a quiet but quick exit.

If you accepted a signing bonus, and you're having to leave before the term is up, you may have to forfeit that bonus.

PRO TIP: Whatever signing bonus you were given, try to negotiate that into your *next* position!

While it may feel emotionally satisfying to "tell off the boss," this is frankly not professional.

It's the sort of stuff that only works in movies or Reddit. In the Real World™, for better or for worse, people will share their personal opinion of you and the circumstances involving your departure. Try to keep things amicable.

Instead of hollering "Bob was a real jerk and I'm out of here!" try to instead say, "I received an offer on a dream project that I simply can't pass up!" If you have managed to secure a new position *before* leaving, as I just hinted strongly that you should, then you'd be telling the truth.

By the way, there's no need to name names, either to your current team or your new one.

Now that we've covered many of the things needed to conduct meetings and build team strengths, it's time to start thinking about what's to come. Check out ***Rule 10: Plan For The Future*** to see what may be your next step in your career.

RULE 10: PLAN FOR THE FUTURE_

Writing code is exciting. Being in the thick of it, bending systems to do your will. It's really cool, but this isn't just some day job.

What we're talking about is a full-fledged career.

And there are things you must not neglect if you want to grow and not be left behind.

THE FIRST STEP in building a lasting career is to not just think about the tickets you're fixing this week. Instead, you must think about where you want to be six months from now, the next year, and the next five years.

This is why it's helpful to create a personal Google Doc to pencil down such ideas.

Maybe today you're fixing a bug on a given project. Is your hope to someday be the lead for that project?

Perhaps there's another project you're not even working on that draws your fancy. Maybe a different team altogether?

So what can you do about it?

One of the most vital things to expand your circle of influence is to seek out bigger tasks and bigger responsibilities. Like we just discussed, maybe today you're fixing bugs. But what if during the next round of project planning, there are some big ideas the team is brainstorming. This may be your chance to pick one of them up!

If you really want to push things, consider a task that isn't already spelled out, but instead would require more thought and effort. This may sound scary. You'd be raising your hand and saying, "I'd like to do that." And while it may make you tremble inside, it could be your most important move to advance your career.

If you're still a bit timid about reaching for such a task, consider the fact that it's a lot harder to become a project lead if you never take on any lead-sized tasks.

Another place to explore goal setting is your immediate manager. Either through an email, a Zoom call, a face-to-face meeting, or your team's instant messenger tool, you can reach out and ask, "What do I need to be doing to become a project lead?"

Notice how I said "to be doing" instead of just "to do?" That's because a simple "to do" sounds like a single task that when done, tees you up for that position. Rarely is advancement a single task. Often, it's a collection of actions carried out over time.

Shortly after my first promotion, I was alerted that part of responsibilities included interviewing new grad candidates. I jumped in and signed up for as many department interviews as I could. But sometimes, we aren't given such a handy heads up. No, sometimes you need to directly ask your manager what is expected of you, whether your current role or your next one.

If you are working in a well-supported team, your manager should have insights into what the higher-ups are looking for.

That's a lot of discussion around the idea of becoming a project lead. However, that's not the only goal you should consider.

That's just an example of thinking about the future of your career.

Maybe today you're a test engineer or a platform engineer, but you want to switch to application development? Or perhaps you want to get involved with your company's developer advocacy team and travel to conferences and user group meetings to demonstrate products? Or perhaps become a field test engineer? The list goes on.

Whatever your goal is, it never hurts to speak with your manager or the leader of your team to express your interests. And it never hurts to seek out tasks and assignments that would align with those goals.

One of my earliest managers would always ask, "what do you want to be when you grow up?" And he was right. In essence, that question is germane along every step of your career.

So take the time to pencil down your thoughts and dreams in that Google Doc I mentioned and try to carve out a vision.

No one else is going to take the initiative to expand your career. That task, along with your daily assignments, is squarely on you.

By the way, I didn't mention it earlier, but an equally valid goal is wanting to get promoted. And it's perfectly okay to ask your manager, "What do I need to be doing to earn a promotion?"

A strong manager will be able to work with you to grow your career. They should be able to coach you on the requirements and what's recommended.

For example, the shop you joined may require a certain number of years as a junior developer before being qualified for promotion to senior engineer. There could be other requirements needed to reach that next level.

Your manager should help you work on various goals that will make you irresistible to promotion by the time you hit that time requirement.

Nothing is guaranteed. And none of these tips will ensure a promotion or a shift to a desired position. Companies are also constrained by market forces and other external factors.

But that doesn't mean you can't work with your manager and team leaders to grow your value to the team and work toward the vision you have for yourself.

Now check out the next section where we dig into a critical thing to grok how your team is evaluating your performance.

So WE'VE DISCUSSED setting goals for your career. A key aspect to achieving those goals is to track your efforts. And the best strategy toward tracking these goals...is to do it yourself.

It's true that communicating your short and long term goals with your manager is vital toward expanding your horizons. But it's also important to realize that it's not up to your manager to track all the details.

Sorry, but that's *your* job.

That's why it is helpful to capture all your goals on a worksheet. And perhaps you can guess what I'm about to suggest? Create either a Google Doc or a Google Sheet. (Yes, I love Google tools. No, this book isn't sponsored by Google!)

After laying out your goals, start tracking your progress on them. This can be a part of your daily routine, or something you catch up on at the end of the week, or perhaps once a month.

The important thing is to start taking notes. If you capture the evidence of meeting goals agreed upon with your manager or team leader, then it becomes much easier to present this to your manager.

You see there is a little something that happens every year among all of your company's managers. You may not be aware of it.

It's vital that you understand this, which is the reason I'm going to pull aside the curtain and reveal a process that is critical for your career. Once you understand, you'll have a much clearer picture of what you need to do.

Every company does what I'm about to share with you, in one form or another. I was once a manager at a previous company, and fellow colleagues I've known from other companies have confirmed it's done in their shops as well. When I was a manager it was called "stacking." Or at least, it was at one time. Other companies may call it something else. But the fundamentals are the same.

All the managers get together once a year and "stack" everyone up from best performance to worst performance. Your manager shows up at this meeting with a list of everyone he or she represents, from their best to their worst team member. And each manager puts up their best candidate.

All the managers will then proceed to work through their stack of reporting employees, and essentially merge these lists together to "stack" the entire team.

Once this is done, they proceed to divvy up any annual merit increases and also assess if anyone is qualified for promotion. Sometimes bonuses are tied to this, sometimes they aren't. This is also an opportunity for the company to dole out any equity grants.

Some shops may grant some of the budget to everyone. Some shops may cut things off at the bottom 10% (or some similar figure) and not grant any bonuses and equity grants to those at the bottom.

Exactly how much bonus money, merit increase, equity grants, and whatever else they have can vary from year to year.

Now at first glance, you may find this procedure detestable. Perhaps it seems a bit...calloused to be stacking up employees like playing cards. Whether you love it or hate it, the concept is quite simple. And understanding it can be your ticket to advancing your career. Let me give you an example.

The last time I participated in this process as a manager, I had a new hire on my team. An individual that carried out any task I extended with flying colors. Someone that I kept giving bigger and bigger scoped assignments. Every stacking meeting I attended, I led with that person.

And every other manager nodded quickly. They knew this person was top-notch. They had either worked with him directly, or one of their reporting employees had. And so it had been no effort to stack this person at the top of the entire department for three years straight.

This person's performance shone through, thus making it easy for me to write the person's promotion packet when the minimum time had passed.

When your manager walks into that meeting, whatever your company calls it, his or her job is to defend you. And what your manager needs is evidence to present to all the other managers. Evidence of your performance.

And this is where you come in.

It's your job to collect the details about your accomplishments. But *not* to impress your manager.

It's evidence your manager can take into that meeting to impress *everyone else.*

And this process cuts both ways. Let me share a personal story.

A somewhat embarrassing story.

Back in late 2021, the Spring Data team had done an

analysis at the behest of upper management as to which modules were the most critical to the community and compared it with how many people were allocated to each project. Spring Data JPA was clearly at the top of the list in community need and usage, but two different people oversaw it. Because these two people also had other responsibilities, no single person was squarely focused on this critical project. And the project suffered because of this.

My manager asked me to stop working on Spring HATEOAS as my primary focus, a project I'd already worked on for four years, and to instead take over Spring Data JPA as my #1 focus.

I was excited to say the least.

I had worked for over ten years in the trenches of relational databases before I had joined the Spring team and could do left outer joins in my sleep. Combined with the fact that Spring Data JPA was very popular, I was stoked to be handed control of this project.

And that's when I discovered that my knowledge of relational databases was great...but my knowledge of JPA was subpar.

I began warming up on the project, but not fast enough. That's the polite way of saying I was in over my head.

For most of 2022, I was struggling to process new tickets. There were several aspects to the project I wasn't aware of. And many of its rich myriad of features I flat out didn't grok.

But that wasn't the worst of it.

I had worked for years as the lead for Spring Web Services with no one else at my side, and this had made my communication skills sloppy. I did a poor job of communicating my inability to effect change with both our team leader as well as my manager.

For 2022, my manager went into the annual meeting with

little evidence of my accomplishments. When he shared with me that he had actually fought on my behalf but had little to fight with, I gulped inside.

That's because I understood the process. I had been there before, doing the job he did. I knew that my lackluster performance had put his own reputation at risk.

He had managed to get me an annual bonus, but very little. And I saw no merit increase.

I had let an entire year slip by without properly communicating my struggles. And I felt like a junior developer just getting started.

He had, in fact, expressed strong support for me at that meeting. Perhaps that is what granted me a second chance?

But instead of falling apart with this feedback, I instead leapt into action.

I wrote down several new goals. I began tracking daily accomplishments (in a Google Doc!). And I vowed to not let another year slip past with no change.

On slow days when I felt overwhelmed, I began sifting through the backlog of Spring Data JPA tickets and found ten-year-old tickets that were frankly never going to happen. I closed them (and logged them!) I began building a record of action and accomplishment.

I also picked up bigger tickets, sought out inputs from my fellow colleagues that had more JPA experience than me, and fixed bugs as well as implemented new features.

Within six months, give or take, my manager was able to share that we had finally put the management team at ease. He even told me that our team leader was wondering where this all came from!

I apologized for my sloppy behavior and instead committed myself to make these short term changes of logging tickets in a work log, of grooming the backlog of tickets, and of circling through current tickets standard procedure.

And I thought it paramount to share this with you, despite my personal embarrassment.

It's possible to get lazy, whether you're a fledgling coder or a seasoned pro. I want you to understand exactly what your manager is doing when he or she has to meet with the other managers and speak on your behalf.

If you find yourself struggling in some area or task, reach out for help!

Open the door with your manager or team leader so you can begin the process of straightening things out. Don't wait as long as I did and suffer monetary loss, or more importantly, putting your reputation at risk.

Now one of the most vital things to boost your career is to continue learning. It's a vital aspect to planning for the future. So check out ***Rule 11: Keep On Learning*** to see ways to never stop learning and increase the value you present to your team.

RULE 11: KEEP ON LEARNING_

New toolkits emerge every week. There are concepts, philosophies, and approaches to coding that weren't here a decade ago. The world of pro coding is constantly evolving.

That's why this chapter is chock full of ways to stay on top of this fun world we call pro coding.

So you finished your CompSci university curriculum? Or perhaps you completed a 12-week boot camp? Either way, you are set, right? No need to learn anything new?

Of course, we both know this just isn't true.

Learning is a continuous process. I once heard a saying, "to become a master, one must always remain a student." Let's check out some ways to keep growing your skill set.

There are lots of places to look for new information. If you haven't already guessed by now, one of my favorite places to learn things is on YouTube. Video is the future.

That's why some of the people I like include:

- Dan Vega (https://www.youtube.com/@DanVega)
- Ken Kousen (https://www.youtube.com/@talesfromthejarside)
- And of course, yours truly (https://www.youtube.com/@ProCoderIO)

The channels listed above are just a sliver of the Java community. What if you're into Python, JavaScript, or Node.js?

Start checking that out! There are *lots* of active developers posting content everyday. Same for React, Python, or whatever toolkit you or your team are using.

It's also worth it to check out the lead developer for a given toolkit to see if they have also written a book.

Books are still quite popular and are getting published all the time.

I'm not talking about reading the latest *Jack Reacher* novel, but instead checking out the next Spring, Python, or JavaScript book. Or whatever other tools you use. Docker, AWS, and Domain-driven design (DDD) are more topics I can imagine you needing to learn that aren't language-specific.

One of the benefits of reading a tech book is that they are usually written by people with much more experience than you. Ideally, you want to find one written by a veritable expert in the field. To kick things off, you can enter your search criteria into Amazon.

For example, if you sought a book on **Spring Boot**, this might be your first page results:

Yes I'm plugging my own book, being the second one on the list. Wouldn't you?

For starters, go and read the description. Then check out the author's credentials. If you want further insight, check out the reviews. Finally, decide whether you favor the paperback or the e-book version.

And not every book will be exactly what you need. Some will be a total bust.

But many will be a source of knowledge for a topic you desperately need to grow in.

And that's worth every penny.

Perhaps buying books feels a little expensive? What if you were new to using Spring Boot, and you got hired onto a team that's using it everyday? That $40 may mean the difference between sinking with failure or swimming with success.

Do that three times a year for perhaps $120 total, and you may equip yourself with knowledge and skills you didn't have, all the while increasing the value you bring to your team.

Remember in ***Rule 10: Plan For The Future*** how your manager uses your skills to defend you regarding merit increases and bonus checks?

Investing in your own skills and the technology you are using is something you can include in your report. Give your manager that extra information about you studying in your field of development.

Or think of it like this. What happens when you *don't* invest in yourself, but a fellow colleague does?

If you bought three books and learned how to leverage the principles of DDD, how to drive Docker containers, and brushed up on your Java skills, your value would rise. If you did none of that, your value may simply stay where it is, but the pro coder next to you who *did*, will rise *relative* to you!

If you're still skittish about putting up that kind of cash, there's a chance your company will reimburse you. Some companies value self-education. And if they don't, ask your manager if you could file an expense report for the books.

But remember, even if you do pay out of pocket, is a $120 investment over a given year really a bad deal?

Of course books aren't the only means to grow your knowledge and skills.

Another avenue of learning involves keeping up with tech news. This means following key individuals. And yes, we're talking about social media!

Social media is a tricky beast. It can suck you in, involving you in the most brutal fights with people you've never met. But it can also be the first place key information is shared.

In essence, find the people that share the content you want. That may sound a bit nebulous, but what I'm suggesting is finding the lead for a toolkit you are using, and see where they post. Perhaps they use X? If you're on X, then you may wish to follow that person. They could be on Mastodon. Do the same. Wherever they are publishing content that is pertinent to your work, consider following them.

As examples, I follow quite a few people in the Spring community, people that write about their latest endeavors on the various projects I'm interested in. You may wish to follow their X handle so you can be alerted to the latest.

The people that develop your IDE may publish valuable content, so consider following them. I happen to use IntelliJ IDEA (https://www.youtube.com/@intellijidea) and they have all kinds of tutorials.

If you are tracking the leaders of the software tools and processes you work with, then checking their postings can keep you more abreast of what's happening. Instead of waiting to hear about it from another colleague at a monthly meeting, why not hear it "straight from the horse's mouth?"

This is when social media serves *you* instead of you serving *it*. This may be a bit challenging, but curating a feed of valuable news and information on the software tools and approaches you work with is quite beneficial.

Another area to grow is by attending a tech conference. As I write this section of the book, I'm flying to the SpringOne conference our team hosts every year. (Well, every year until the lockdowns in 2020). It's an exciting opportunity to meet and learn new ideas. If you're a Spring/Java coder, then check out the next opportunity to attend!

If your team is using Python, try to find a conference near you. By simply looking for "top python conferences" on Google, I found eleven different events for 2023, hosted in France, Canada, United States, Spain, Germany, Italy, UK, and Czech Republic. Many are hosted virtually!

As a secret trick, ask your manager if your company would send you there. They may be willing to pay the fee of the conference, cost of the hotel, and if you have to travel, perhaps airfare? Remember, this is an opportunity to grow your skill set and serve your team, so pitch it to your manager like that!

There is yet another reason to consider going to a conference *even if* you must pay for it yourself. A developer conference where you spend 2-5 days rubbing elbows with fellow coders talking about tech and countless others things will *recharge you.*

I have spoken at conferences since 2008, and after every one, I've been imbued with new energy and incredible excitement. This type of self-investment can also fill you with fresh ideas and possibly give you the inspiration to carry them out.

As exciting as conferences are, we can never find nor afford to go to one every month. Hence, another resource to invest in your own personal pro coder growth, is to find a local user group. The Java community has Java User Groups (JUGs). I personally launched the Nashville Java User Group back in 2010, and finally turned it over to a friend in 2023. And in that time frame, was able to meet and get to know several developers in the area. A huge win for myself and for everyone that became part of that group.

You may be able to find some sort of user group through Meetup.com, Facebook, X, or elsewhere. The first step is to start looking. With the amazing power of Google, simply enter "<your city> <your tech> user group" and then check out the results. Below, you can see "nashville react js user group":

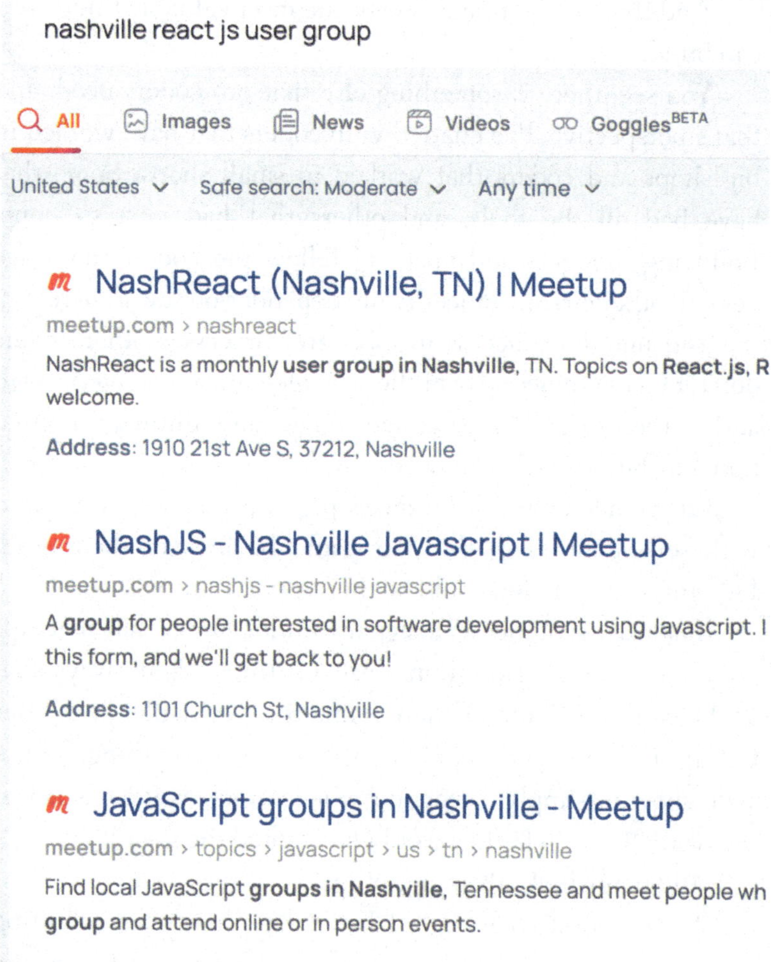

See? Plenty of opportunities!

And for the record, I have never seen a programming-oriented meetup group that requires paying any dues to attend their meetings. Many even offer free food!

See when their next meetup is...and go! It's not only a chance to learn more about the tech you're using, it's a valuable opportunity to network with fellow developers in your region.

And that connection is one of the most valuable things you can build.

You see, there is something else that pro coders need, and that's perspective. I've chatted with coders that have worked in big shops and coders that worked in small shops. Some that have had all the tools, and others that had next to none. Bouncing thoughts and ideas off fellow pro coders that have been in other circumstances really expands your awareness.

You may have access to tools and processes other coders don't. Having someone hear the struggles you are going through and to then say, "That sounded rough" and affirm your situation may be the thing you need.

Or...someone at a local meetup group that you've chatted with several times in the past could be in over their head. Hearing your reaction could be the support *they* seek.

That last sentence is a signal to understand that meetups aren't just for consumption. Instead, think about ways that YOU can contribute. When I ran the Nashville Java User Group, I encouraged people that if they had anything, be it a slide show or simply a reaction to some tool, as long as it had the word "Java" in it, they would be granted the stage.

And many took advantage of that.

At one point, years ago, I had to step away from hosting meetings for a few months, for personal reasons. I feared the group would fall apart, but to my surprise, it actually grew the most at that time. I had never seen it with more than a dozen people up to that point, but when I returned, there were over 60 people at a meeting!

Not sure if that reflects on me. ;)

In all seriousness, the group rose up to that challenge and hosted meetings without me, which is actually a good thing.

The point is, find a local group, connect with them, build relationships with them, learn from them, and then be willing to contribute as well. Don't simply take, but think about producing.

Hopefully, these various ideas serve as inspiration on ways you can grow your own knowledge, experience, and professional network as well.

Now if you start thinking about presenting at local user groups, perhaps you'll consider becoming a speaker at a regional or national conference. There are some extra things you need to consider doing if that's the case.

IF THERE's one thing that's constant in software development, it's the fact that things are always changing.

We certainly aren't using the same browsers, work machines, coding tools, or toolkits we were using even a decade ago.

One keen way to stay on top of trends is by reading blog posts and technical articles. In this day and age, there are a wealth of resources.

Various authors are posting articles on Medium or their own website. Feel free to check out my own Medium site at https://procoderio.medium.com/. My content contains lessons every pro coder should learn!

LinkedIn has also become an increasingly popular platform for tech leaders to post their content. This is yet another reason I mentioned back in ***Rule 1: Curate a Professional Persona*** to establish your presence on LinkedIn.

Reading other people's articles, posing questions in the comments, and interacting with fellow coders is quite valuable. Not only is it a great platform to tap into with focused content in your field. It becomes a place to build valuable relationships on a professional level.

Some people will write articles directly on LinkedIn, while others may simply post links. Either way, it's handy for you.

There are other ways to find technical articles. In the past, I used Google News Reader to monitor about a hundred various blog sites through the power of RSS. Anytime someone posted a new article, it showed up as a new article on my dashboard. So once a day, I could check in and catch the latest.

Now let me be clear, I'm talking about tracking tech blogs not general "news."

Unfortunately, Google shutdown their news reader app years ago. It was further destroyed by the rise of social media networks like X and Facebook.

So in a sense, your best bet to find tech news is to get on X and follow the people that are writing articles and sharing their links on that platform.

As for Facebook, in general I *don't* recommend them. They tend to be less favorable to tech content. Anything you find is typically gone by the end of the day, never to be found again. It's too much scrolling for too little in value.

However, another platform that may be more content rich is YouTube.

There are an ever growing number of content creators producing tech-based videos. And I'm not talking about iPhone reviews, though those exist.

No, I'm talking about software coders making videos about using your favorite tools. In the space of Spring alone, you'll find people like Dan Vega, Maciej Walkowiak, and of course, yours truly.

Video is one of the fastest growing mediums for, well, anything. And professional coding is no exception. That's why I suggest you find some channels that cover the tech you use and subscribe to them.

People are also posting video content on LinkedIn, so it's another great way to beef up your skills.

This whole message of getting on social media is to seek out content that serves your professional needs.

Now I'm not saying that you need to be there everyday.

But finding a new article to read or a new video to watch once a week will grow your knowledge.

And while social media is of value, there is another website that you should spend extra attention on: StackOverflow.com.

Stack Overflow is a Q&A website where people from around the globe pose questions on various programming languages and software toolkits. There's a good chance that anything you're using, be it Python, TypeScript, Go, Spring, Docker, Postgres, Rust, or whatever, you'll find questions.

In fact, some people have expressed the utmost respect for the maintainers of Stack Overflow that are forced to fix it when the site is down. After all, where are they supposed to find their answers?

But there's another hot area for technical content—podcasts.

Podcasts are a happening place.

Just about every niche, every technology, and almost every team has some variant of audio podcasts, video live streams, and other such "modern" approaches. Whether it's a developer advocate hosting it (check out Spring Office Hours hosted by Dan Vega and DaShaun Carter at https://procoder.io/spring-office-hours as just one example), an engineer from a particular team, or a collection of like-minded developers from multiple teams, there's always another podcast to tune into.

Now we've mentioned all kinds of places where you can consume technical content to grow your skills. But have you built up your own *personal* brand? Read the next section to learn more.

So YOU'RE THINKING about not just scooping up videos and presentations at user groups and conferences, but instead actually speaking? Maybe you're also interested in expanding your network? No, I'm *not* talking about the number of people you're connected with on LinkedIn!

There is something you should seriously think about known as your **personal brand**. What exactly is a personal brand?

Think of it like this, when you get up in front of a group of people, what do you want them to see and hear? How do you want them to see *you*?

We've talked about this in previous chapters in different ways, when we discussed things like X, YouTube, and other social media locations. This is also true if you want to go down the road of blogging content.

Presenting a talk at a user group or a conference is simply an extension of that. All these mechanisms through which you produce content to share with others can all be random, disjointed, and purely chaotic.

Or...you could consider having a unified theme or motif. Let me share an example.

Several years ago, I noticed a colleague of mine update his team profile with the moniker "JavaScript Barbarian."

I laughed. Hard.

It was a phrase that made me pause and think. This person was brilliant at JavaScript to the point that he'd once run a consulting business.

It was a way for him to brag about his talent without sounding condescending or rude. And then I spotted him putting the same description on his Twitter bio. He had adopted this "title" and ensured that if you spotted him on any of the places he had staked out in social media, you knew it was him. Since I was studying JavaScript a little under him, I considered myself a "JavaScript Padawan" at the time!

For years, I had called myself a "test-bitten script junky", and so I immediately updated my own X bio and team description. A few more iterations over the next couple of years, and I'd honed my "thing" for others to see.

Check out my latest incarnation below:

You can see my description, the biggest things I want people on this platform to know me for, and a little bit of personality. Overall, I'm trying to give you a bit of a vibe of who I am and what I do.

Anytime I have a chance to speak, whether it's to someone in an elevator at a conference or up on stage, I think about how I want to tie into my personal brand.

You don't always have to do this. Don't be pitching every

person you bump into. But to hone a personal brand, you want to strive for a message that you believe in, and then seek to include it every chance you have to spread your message.

You may not realize it, but this type of focus on your own message can make you begin to look like a senior developer, not a junior dev that's just getting started. Junior developers are scrambling to learn the ropes of writing code. Senior developers are making waves.

Which one do *you* want to be?

Mind you, a personal brand can take years to grow. I'm not talking about the number of followers or subscribers on social media. Instead, this is uncovering what's important to *you* such that you're *excited* to share it in the venues you visit.

That's the part that can take time.

What do you see in the world of pro coding that's valuable that you wish to impart to others?

To fully round things out, I had always enjoyed sharing the latest Spring coding tech content. But in the past couple of years, I noticed that certain things had a limited shelf life. Spring books aged faster and faster, forcing me to write updates quicker.

But other things are more fundamental, more eternal. It's one of the defining reasons I decided to write this book. It's okay to pour into immediate things, but also consider things that are more evergreen as well.

Learn what excites you and what you want to share with others, and pursue that! Which leads us to the last chapter, perhaps the most important one: ***Rule 12: Help Others***.

RULE 12: HELP OTHERS_

In ***Rule 11: Keep On Learning***, we talked about various ways to perpetuate your growth. Toward the end, we discussed building your personal brand. But there is something that building your brand is tantamount for. Something I feel is a vital duty we all should strive for—helping others.

I believe we are given a gift to learn and build systems. But even more importantly, we have the chance to help the next generation of software developers.

Have others helped you along the way? Have others created books, articles, tutorials, and videos?

One of the things we should pursue is helping others. Not just those using our systems, but also our fellow colleagues and community members that are also building systems.

In this chapter, you'll learn many ways to better help your teammates, community members, and others.

ONE OF THE most prized things any of us can get is feedback. While we discussed ways to request feedback from fellow colleagues through pull requests in ***Rule 8: Manage Your Tickets***, the ever important flip side is offering solid feedback to others.

For starters, does this code uphold the standards your team has agreed to? Some teams may be a little lax and offer much flexibility, but some teams (e.g. the Spring team) have very specific styling they prefer in their code. If you are making comments on someone's pull request, by all means feel free to point it out if you spot something amiss.

This includes typos or improper casing. For example, in Java it's common to use lazy camel case in the naming of fields. If you see **AccountNumber** when it should really be **accountNumber**, then all you have to do is add a comment that says, "I think this should be accountNumber."

You may wish to check your team's policy on the usage of acronyms for names. For example, it's valuable to know if your team prefers **HTTPMessageConverter** or **HttpMessageConverter**. Assuming you've gleaned that, you can now mark up your colleague's pull request if they aren't upholding the standard.

All the things mentioned so far are what I like to think of as the *easy stuff*. There isn't a context I can imagine where a typo in a field name would be appropriate. And calling out lack of adherence to team standards is always useful. So it's nice to read through the suggested change and offer comments that are cut and dry.

But that isn't the only form of feedback your colleague may be seeking.

Is this person's design appropriate? This is where you start moving away from straightforward facts and instead toward *opinions* on how to effect change.

Perhaps this pull request shows the user introducing a Java enum? Java enums definitely have benefits as well as downsides. In this pull request do the benefits outweigh the downsides or not? If you see a problem with the proposed change, you can share that opinion.

What if you aren't convinced of the proposed change, but you also aren't quite sure why? It's also possible to ask questions and consider that feedback. "I'm not sure why you are introducing a new interface. Could you expound upon your reasoning a little?"

Simply asking questions and furthering the conversation is acceptable.

Now let's pause for a moment and double check a critical philosophy all pro coders should consider. All of your efforts to correct typos, uphold standards, or to discuss new designs will be for naught if you don't adhere to The Pro Coder's Motto: **Be Professional**. What does this motto mean, exactly?

This is more than being courteous and kind.

No.

To **Be Professional**, you must do much more than that. You must wake up, go to work, and trust in your colleagues. Extend grace and consideration. If someone is giving YOU feedback that grates on your nerves, pause and give them due consideration. It's probably not *personal*, but instead *professional*.

Give them the benefit of the doubt.

And when you give feedback to others, try to put it through the same lens. Don't demonstrate attitude, but instead compassion. Someone that is showing you a pull request which touches twenty files probably spent several hours working on that feature.

Take the time to read their code. Decide upon your opinion about their efforts. And then work to express your opinion in a *professional* fashion. It's actually quite remarkable if you slow down enough to compose your opinion in your head before blurting it out, whether that's face-to-face in a meeting or as a comment on a GitHub ticket or pull request!

But it doesn't stop there. There is one *more* facet I must share that is not so easy.

Nine times out of ten, your teammates and you will find that being professional isn't that hard. But there will be that one instance, that moment when a colleague may ping you on Slack (or whatever tool you all use to message each other) and lose their cool. Your co-worker may be quite upset over something.

Perhaps you made a major mistake? Perhaps not. Maybe you made an assumption and took action that wasn't good for them. Whatever it is, this is the moment where you must decide if you truly *are* a pro coder.

Because the emotional desire to respond in kind will be overwhelming.

Instead of shooting from the hip and rapidly defending your actions, just stop.

Listen.

Try to understand why your teammate is upset. In fact, when it's your turn to respond, simply try to reiterate *why* they are upset. Repeat back to them what *they* appear to be saying has gone wrong. Also, do it with a tone of voice that shows you are sincerely trying to grasp it from their perspective.

You'll be amazed at how this simple communication tool can defuse situations and let you focus on solutions.

This is what it means to **Be Professional**. By putting aside your own opinion, for just a moment, and making their perspective the priority, you will build a professional rapport of lasting significance.

Which leads us to the next section where we discuss ways to foster a positive work environment.

WE TALKED about The Pro Coder's Mantra, **Be Professional** and what it means to carry that out. More than anything, it's a philosophy not just about coding, but about interacting with fellow co-workers. In our ever-evolving tech world, things move fast.

Opinions that were right at one point in time, may become invalid before a year has even passed. Dogmatic adherence to rules or technical decisions leaves no room for growth and nuance. However, **being professional** with fellow colleagues builds lasting relationships that makes it possible to grow and expand and ultimately achieve amazing things.

In September 2023, I attended the SpringOne conference, the first in-person SpringOne since the COVID lockdowns. At that event there weren't just fellow co-workers. There were members of our community from around the world. And there were past co-workers that had moved on to other endeavors. The fact that we had spent time building professional relationships made it easy for me to respond to a previous co-worker's request, "Greg, would you mind getting together? I'd really like to discuss something!"

The fact that this previous co-worker and I had built up a professional connection made me want to drop everything and help them out.

Someone was asking *me* for help with something, and so I sent him my cell phone number so he could text me. I didn't know what he wanted, but I would not let tech or the fact that he lives in a different country be the reason I wasn't able to try and serve his needs.

This type of commitment to help fellow professionals, even if the words are never spoken, is never forgotten.

Many of us at that SpringOne conference also staffed our

team booth. Spring users would come by, and we'd ask if they had questions or were running into problems. Making yourself available to others, whether they are allies or competitors, is sometimes known as "coopetition." It's a value I believe in and have never manifested more strongly than in software development.

Amidst all these companies, we all work on different products. Some are competitive, some are not. But by upholding a commitment to **Be Professional**, we can take the time to listen to what others have to say and choose how best to serve.

There are some other ways you can reach out to help others.

THERE'S A SAYING that you don't really understand something until you teach it to others. StackOverflow.com is the place where helping others solve their problems can actually help you grow your own knowledge set.

Consider visiting the site once a week, and finding a particular category or "tag" and trying to answer a question. And then next week. And then the week after that.

I actually know someone that committed to answering a question a week about a specific Spring project, for more than a year. An opportunity opened up to join the Spring team, and in the midst of the interview process, the hiring manager said, "We already saw all your answers on Stack Overflow. You've pretty much earned the position."

That person didn't succeed because they took a shot. No, they had clearly learned the details to give solid answers, and built up a lengthy trail, proving not only their own knowledge, but also demonstrating incredible community support.

What are you doing to build such a reputation? By sharing

what you have learned, not only can you help others, but you can also accrue a record of having done so. Check out my own dashboard!

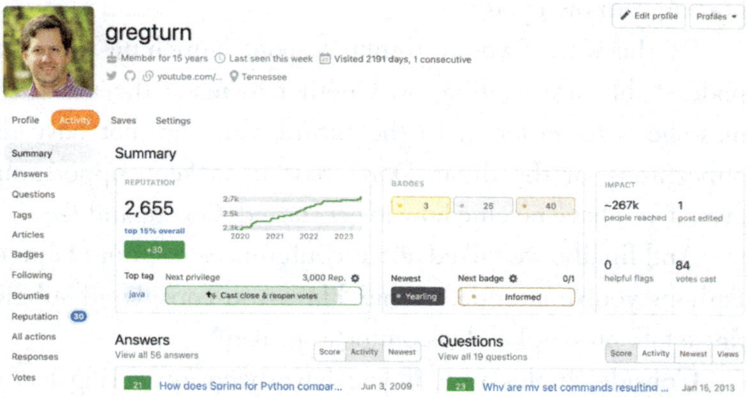

This may not be the very best reputation, but it's certainly more than zero.

Now this isn't about gaming the system. This is actually going out and sharing what you have learned with others.

There's a certain thing that happens when you move from learning the technology to teaching others. Lessons taught also teach and inform the teacher!

Instead of simply consuming content, become part of the next generation that is producing it, in one form or another. You may be a fledgling coder today, but one day in the future, as your knowledge expands and your wisdom deepens, you will have opportunities to share with others.

Frankly, I believe it's a solemn duty that we strive to share what we have learned through any means possible! It almost feels selfish to have learned the ins and outs of this industry, and to *not* share it with others. This is a strong part of my motivation in writing this book.

I've already mentioned in prior chapters things like reading books, catching up on tech articles on Medium and other blog sites, and I've talked about answering questions on StackOver flow.com.

Another hot place to produce content is YouTube as well as hosting your own podcast.

By the way, if you're thinking about launching your own podcast, but are waffling on whether to make the move, my message is to *go for it*. In the future, you may not have the opportunity or the drive. Don't pass up a keen opportunity even if you have no clue how to get started. Go out and *learn*.

And finally, we talked about conferences in prior chapters. Perhaps you've attended some? Have you considered submitting a talk proposal and becoming a speaker?

Consider it this way, if you have been answering lots of questions on Stack Overflow and have become really well versed in a particular toolkit, try to hunt down a relevant conference that covers that tech. For example, if you use Angular and have become quite adept at it, start scoping out **ng-conf**, probably the most popular conference for the Angular community.

Put together at least three proposals, and be sure to send them in when the call for papers (CFP) opens. I always suggest a minimum of three talks, because some conferences want to minimize costs. If they can get three talks out of you and only have to give up one ticket to do it, it's much better for them.

Some conferences will not only cover the cost of the conference, but may extend coverage for travel costs and lodging. Others won't. Some conferences count on the speaker's company to cover said costs. Do a little research on what a given conference covers, and whether or not your company would be willing to cover the difference. That way, it's not costing you anything.

And if you make it onto the speaker's list, that's one more bullet point to add to your LinkedIn profile!

But seriously, if you earn a place as a speaker at a conference, it's a signal that you should seriously look at your speaking options every year from there on. Some people actually earn significant returns speaking at multiple events. I happen to know a consultant that speaks at multiple events. Getting his name out there even earns phone calls for new clients.

Something to realize is that every version of content that you consume also has a way for you to produce. And if you're the producer, it puts you in the driver's seat. People that write books, write blog articles, answer questions, host podcasts, create videos, or speak at conferences are of much higher value in the eyes of hiring managers.

Imagine walking into an interview for a new position and being able to show them multiple articles you wrote about TypeScript while also having spoken three times at various TypeScript conferences? You are signaling that if they hire you, they aren't just getting a software developer. They're getting someone that could generate content for the company!

In summary, your "day job" may be writing software. But writing blog articles, answering stackoverflow.com questions, engaging in interviews, joining live streams, and crafting content on LinkedIn are part of what a true pro coder does. Just be sure to coordinate your efforts with your manager.

You may be just starting out your career and getting the hang of coming in everyday and crafting the code. But at a certain point, probably earlier than you realize, simply writing code isn't enough to satisfy the demands of being a pro coder.

Don't simply assume your job is to sling code from 9-to-5, go home, and chill out on the couch.

It's not.

As you attain experience, your proportion of time spent writing code will actually go down. It may never completely disappear (unless you join the dark side, a.k.a. management!), but your company will certainly expect you to rise up and give presentations, carrying the banner of your team and your products at various user groups among other things.

If you're skeptical about this, simply ask your manager in your next 1-on-1 what they would think if you wrote an article for InfoQ? You might be surprised!

On a side note, I launched my own YouTube channel back in 2019 with the plans to make videos about Spring Boot, which has since evolved to how to become a pro coder. Since then, every single platform I've seen (Facebook, LinkedIn, X, Instagram, etc.) have given a *strong* signal that they want people creating video content. In all honesty, I believe video is the future.

You don't have to make videos. After all, that's quite an undertaking.

No, I'm simply suggesting that you could craft your own blog site using Medium.com, contact InfoQ.com, or launch one using your own GitHub account. There are plenty of static site-builders for GitHub. Or you could create a Wordpress site.

The point is, that as a pro coder, as you learn more and gather your own wisdom, the rest of the software world *needs* you. We need your voice and your own slant on how things come together.

In fact I feel this to be a personal responsibility for seasoned coders.

Maybe you're not ready today? That's fine. But someday you may feel a drive to launch your own content. Well here I am, giving you full authority to proceed! Don't forget to let me know when you kick things off.

And this draws things to a close. Here in ***12 Rules for***

Pro Coders, I started with a collection of tips to help you become a savvy pro coder. It's only fitting that I wrap things up by ensuring that you pay it forward in turn by helping others also become savvy pro coders. Be sure to tell those in your circle to also **Be Professional**. And if you pledge to uphold the Pro Coder's Motto to **Be Professional**, then visit https://www.procoder.io/be-professional and grab a shirt or sticker!

If you enjoyed this, be sure to visit https://procoder.io/12rules and sign up for yet more content. This most certainly won't be my last book, and for anyone that has made it to the end, you are the first one I will be contacting to share it in the future.

Thank you for your time.

—Greg L. Turnquist

ABOUT THE AUTHOR_

Greg L. Turnquist worked on the Spring team for over thirteen years and is a senior staff technical content engineer at Cockroach Labs. He was the lead for Spring Data JPA, Spring Web Services, and a committer to Spring HATEOAS, Spring Data, Spring Boot, and R2DBC. He wrote Packt's best-selling title, *Learning Spring Boot 2.0 2nd Edition*, and its *3rd Edition* follow-up as well as *Hacking with Spring Boot 2.3: Reactive Edition* and other titles. He co-founded the Nashville Java User Group in 2010 and hasn't met a Java app (yet) that he doesn't like.

Signup at **procoder.io/12rules** and catch all of Greg's future works. You'll get a FREE technical handout as a bonus!

Be sure to subscribe to Pro Coder, the YouTube channel that helps you accelerate your coding career with actionable advice at **youtube.com/@ProCoderIO**.

ALSO BY GREG L. TURNQUIST_

Enjoy his other published works (**https://trnq.st/books**):

What Is A Database?

Learning Spring Boot 3.0 3rd Edition

Hacking with Spring Boot 2.4: Classic Edition

Hacking with Spring Boot 2.3: Reactive Edition

Learning Spring Boot 2.0 2nd Edition

Learning Spring Boot

Python Testing Cookbook

Spring Python 1.1

Made in United States
Orlando, FL
04 November 2026

90207809R00115